Essays in
Biochemistry

Essays in Biochemistry

Edited for The Biochemical Society by

P. N. Campbell

Courtauld Institute of Biochemistry
The Middlesex Hospital
Medical School
London W1P 7PN
England

R. D. Marshall

University of Strathclyde
The Todd Centre
31 Taylor Street
Glasgow G4 ONR
Scotland

Volume 16

1980

Published for The Biochemical Society by Academic Press
London, New York, Toronto, Sydney and San Francisco

ACADEMIC PRESS INC. (LONDON) LTD
24/28 Oval Road
London NW1

U.S. Edition published by
ACADEMIC PRESS INC.
111 Fifth Avenue
New York
New York 10003

British Library Cataloguing in Publication Data
Essays in biochemistry.
  Vol. 16
  1. Biological chemistry
  I. Campbell, Peter Nelson.   II. Marshall, R. D.
  574.1 '92   QH345   65-1522
ISBN 0-12-158116-0

Printed in Great Britain by
Spottiswoode Ballantyne Ltd.
Colchester and London

Biography

Edward J. Wood received his D.Phil. in 1971 from the University of Oxford. He has worked for The Wellcome Foundation in Beckenham, Kent, and for a spell was a lecturer in Biochemistry in the Medical School of the Royal University of Malta. He is now Senior Lecturer in the Department of Biochemistry, University of Leeds, and is mostly interested in gastropod respiratory proteins.

Professor Brenda E. Ryman graduated from Cambridge during the War, and after a spell at Glaxo Laboratories working in the research labs was seconded to Birmingham University. She eventually returned there to study under Professor Thorpe in the Medical School for the Ph.D. degree. Medical schools have been her main working place—the Royal Free Hospital Medical School from 1948–1972 (where under Professor W. J. Whelan she developed a strong interest in glycogen metabolism and its disorders), and Charing Cross Hospital Medical School from 1972 until the present, apart from study leave in Imperial College for a year in Sir Ernst Chain's department. From 1976 she has also been Mistress of Girton College, Cambridge, and has managed (not always very well, she says) to combine a career with marriage and a family.

David A. Tyrrell graduated from Queen Elizabeth College (University of London) in 1973. He proceeded to a Ph.D. in 1976 working in the Department of Biochemistry, Charing Cross Hospital Medical School, where he remained for a further two years—including six months in the U.S.A.—until joining the Radiochemical Centre, Amersham, in 1978.

James F. Tait graduated and took his Ph.D. in Physics at the University of Leeds. In 1948 he became a Lecturer in Medical Physics at the Middlesex Hospital Medical School. From 1958 he did research in endocrinology at the Worcester Foundation for Experimental Biology, Massachusetts, U.S.A. and then returned to this country as Joel Professor of Physics as Applied to Medicine at the Middlesex Hospital Medical School in 1970. His direct research interests have been in biophysical methodology, characterization of aldosterone, steroid dynamics and the dispersed cells of the adrenal cortex.

Sylvia A. S. Tait graduated in Zoology at University College, London. She joined the Courtauld Institute of Biochemistry at Middlesex Hospital Medical School working mainly on the biological activities of natural and synthetic oestrogens and later on the isolation of aldosterone. She continued in endocrine research at the Worcester Foundation for Experimental Biology, Massachusetts, U.S.A. and returned to the Middlesex Hospital Medical School in 1970 as a Research Associate on MRC Programme Grant. Her major field

v

of interest has been the biosynthesis and mechanism of control of steroid production by cells of the adrenal cortex.

Janet B. G. Bell graduated in Physiology and Biochemistry and obtained her Ph.D. at the University of Southampton. After lecturing at Brunel University, she spent several years as a Senior Research Fellow in the Department of Zoology, St. Bartholomew's Medical College studying the testicular steroidogenesis of man and other mammals. She is now working with Professor and Dr. Tait in the Biophysical Endocrinology Unit of the Physics Department, Middlesex Hospital Medical School, on the steroid metabolism of the different zones of the mammalian adrenal cortex.

Preface

Kind Sir, I've read your paper through,
And faith, to me 'twas really new!
ROBERT BURNS (1759–1796)

The unprecedented rate of growth in scientific effort, after the Second World War, encompassed major developments in such diverse areas involving biological phenomena as medicine, agriculture, microbiology, food science, environmental science and technology. The resulting growth in the body of knowledge of the chemistry and physics of the life sciences in general led quite naturally to a mounting literature both in the discipline of biochemistry itself and in a variety of other specialized areas.

The student and teacher of biochemistry became faced with the need to read increasing numbers of papers in more and more highly specialized areas. Some of the papers were "really new" in that they pioneered either new ideas or new experimental procedures, whilst many were concerned with testing the ideas or with further development of technique. Although specialist reviews of the various developing areas rapidly became available, the Committee of the Biochemical Society saw that there was a need for the advanced student to have available to him a series of Essays, setting out biochemical topics in a stimulating manner, which showed the background, the present state and possible future developments in the particular area, and which could moreover be read with pleasure and profit. It was from these thoughts that Essays in Biochemistry became a reality with the publication in 1965 of the first volume under the editorship of Peter Campbell and G. D. Greville.

The past fifteen years have seen a further expansion of the subject of biochemistry into all areas of biological sciences and the topics in the present volume indicate the diverse nature of the problems which the biochemist seeks to resolve. Haemoglobin and the haemoglobinopathies have been discussed extensively elsewhere and the Essay by E. J. Wood discusses the less well developed area of the ways in which oxygen is transported and stored by invertebrates. Further work in this area may lead to a better understanding of evolution. The Essay concerned with liposomes (B. E. Ryman and D. A. Tyrrell) stresses the potential uses, yet unresolved problems, in medicine, but at the same time the excitement of a topic developed in an interdisciplinary manner by, among others, physical chemists, pharmacologists and clinicians as well as by biochemists becomes evident. The more specialized nature of the

vii

Essay on Steroid Hormone Production by Mammalian Adrenicortical Dispersed Cells (J. F. Tait, S. A. S. Tait and J. B. G. Bell) not only shows some of the ways in which the problems of steroid production are being tackled, but it also emphasizes some of the general aspects likely to be encountered when studies are made or cells removed from their normal pericellular environment, and the need to give great attention to the effects of the latter. Each of the topics has implications in other areas of study. We hope that this group of Essays will indeed be read with pleasure and profit.

This volume of Essays is only the second in which the writer has been involved as an Editor and it was only after considerable hesitation that the invitation to act as co-editor of the series was accepted. Essays in Biochemistry has achieved high standards under the editorship of Peter Campbell, with G. D. Greville, Frank Dickens and Norman Aldridge successively. Some degree of apprehension was naturally felt concerning the need to help maintain these standards and at the same time concerning the requirement for gaining the co-operation of authors who are likely to have, invariably, considerable individuality of thought. The second of the apprehensions was rapidly dispelled by the realization of how much time and effort is given by authors in describing their subjects in stimulating ways, and for this we are grateful. Encouragement was provided also by a number of individuals who indicated how much Essays was valued. The series is more likely to continue as a successful venture if the reader will suggest suitable topics for future volumes and will also offer constructive criticisms to the Editors. These will be welcomed and carefully considered.

August 1980   R. D. MARSHALL

Conventions

The abbreviations, conventions and symbols used in these Essays are those specified by the Editorial Board of *The Biochemical Journal* in *Policy of the Journal and Instructions to Authors* (revised 1976 *Biochem J.* **153**, 1–21 and amended 1978 *Biochem. J.* **169**, 1–27). The following abbreviations of compounds, etc., are allowed without definition in the text.

ADP, CDP, GDP, IDP, UDP, XDP, dTDP: 5'-pyrophosphates of adenosine, cytidine, guanosine, inosine, uridine, xanthosine and thymidine

AMP, etc.: adenosine 5'-phosphate, etc.

ATP, etc.: adenosine 5'-triphosphate, etc.

CM-cellulose: carboxymethylcellulose

CoA and acyl-CoA: coenzyme A and its acyl derivatives

Cyclic AMP etc.: adenosine 3' : 5'-cyclic phosphate etc.

DEAE-cellulose: diethylaminoethylcellulose

DNA: deoxyribonucleic acid

Dnp-: 2,4-dinitrophenyl-

Dns-: 5-dimethylaminonaphthalene-1-sulphonyl-

EDTA: ethylenediaminetetra-acetate

FAD: flavin-adenine dinucleotide

FMN: flavin mononucleotide

GSH, GSSG: glutathione, reduced and oxidized

NAD: nicotinamide-adenine dinucleotide

NADP: nicotinamide-adenine dinucleotide phosphate

NMN: nicotinamide mononucleotide

P_i, PP_i: orthophosphate, pyrophosphate

RNA: ribonucleic acid (see overleaf)

TEAE-cellulose: triethylaminoethylcellulose

tris: 2-amino-2-hydroxymethylpropane-1,3-diol

The combination NAD^+, NADH is preferred.

The following abbreviations for amino acids and sugars, for use only in presenting sequences and in Tables and Figures, are also allowed without definition.

Amino acids

Ala: alanine	Asx: aspartic acid or	Cys or Cys: Cystine (half)
Arg: arginine	asparagine (undefined)	
Asn*: asparagine	Cys: Cysteine	Gln†: glutamine
Asp: aspartic acid		Glu: glutamic acid

* Alternative, Asp(NH₂)   † Alternative, Glu(NH₂)

Glx: glutamic acid or
  glutamine (undefined)
Gly: glycine
His: histidine
Hyl: hydroxylysine
Hyp: hydroxyproline

Ile: isoleucine
Leu: leucine
Lys: lysine
Met: methionine
Orn: ornithine
Phe: phenylalanine

Pro: proline
Ser: serine
Thr: threonine
Trp: tryptophan
Tyr: tyrosine
Val: valine

Sugars

Ara: arabinose
dRib: 2-deoxyribose
Fru: fructose
Fuc: fucose
Gal: galactose

Glc*: glucose
Man: mannose
Rib: ribose
Xyl: xylose

* Where unambiguous, G may be used.

Abbreviations for nucleic acid used in these essays are:

mRNA: messenger RNA
nRNA: nuclear RNA
rRNA: ribosomal RNA
tRNA: transfer RNA

Other abbreviations are given on the first page of the text.

References are given in the form used in *The Biochemical Journal*, the last as well as the first page of each article being cited and, in addition, the title. Titles of journals are abbreviated in accordance with the system employed in the *Chemical Abstracts Service Source Index* (1969) and its Quarterly Supplement (American Chemical Society).

Enzyme Nomenclature

At the first mention of each enzyme in each Essay there is given, whenever possible, the number assigned to it in *Enzyme Nomenclature: Recommendations (1972) of the International Union of Biochemistry on the Nomenclature and Classification of Enzymes, together with their Units and the Symbols of Enzyme Kinetics*, Elsevier Publishing Co., Amsterdam, London and New York, 1973: this document also appeared earlier as Vol. 13 (2nd edn, 1965) of *Comprehensive Biochemistry* (Florkin, M. & Stotz, E. H., eds), Elsevier Publishing Co., Amsterdam, London and New York. Enzyme numbers are given in the form EC 1.2.3.4. The names used by authors of the Essays are not necessarily those recommended by the International Union of Biochemistry.

Contents

The Oxygen Transport and Storage Proteins of Invertebrates

E. J. WOOD

Department of Biochemistry, University of Leeds, 9 Hyde Terrace, Leeds LS2 9LS, England

I. Introduction

During the course of evolution animals have developed a number of ways of transporting and storing oxygen, one of them being, in vertebrates, possession of a tetrameric haemoglobin in circulating corpuscles for oxygen transportation and a monomeric myoglobin in muscles for storage. A range of oxygen-binding proteins is found in invertebrate animals, many having very beautiful and complex architecture, and all having striking colours because of the presence of transition metal ions at the oxygen-binding site (Table 1).

Probably because of their colours, these proteins have long attracted the interest of biochemists and physiologists, and they had a very significant influence in leading Svedberg to propose that proteins had discrete molecular weights and that protein molecular weights were simple multiples of a

1

fundamental unit.[1] These ideas were important for the development of our understanding of protein quaternary structure.

The oxygen-binding centres of the respiratory proteins can involve either iron–porphyrin, iron, or copper (Table 1). It is doubtful whether any respiratory pigments with completely different centres remain to be discovered. Thus although the blood cells of tunicates contain vanadium (as "haemovanadin") it seems almost certain that this is not a protein complex,[2] nor is it capable of binding oxygen reversibly.[3]

TABLE 1

The oxygen-binding proteins of animals

Name	Oxygen-binding centre	Stoichiometry	Colour change oxy/deoxy	Mol. wt. per O_2 bound
Haemoglobin, myoglobin	Protoporphyrin IX–Fe(II)	$Fe:O_2$	Red/ red-purple	17 000
Invertebrate haemoglobin (erythrocruorin)	Protoporphyrin IX–Fe(II)	$Fe:O_2$	Red/ red-purple	17 000 (?)
Chlorocruorin	"Chlorohaem"†	$Fe:O_2$	Green/red	17 000 (?)
Haemerythrin	Non-haem Fe(III)	$2Fe:O_2$	Burgundy/ colourless	13 500
Haemocyanin	Non-haem Cu(II)	$2Cu:O_2$	Blue/ colourless	50 000 (molluscs) 75 000 (arthropods)

† Protoporphyrin IX in which a formyl group is substituted for a vinyl group at position 2.

In this review I have tried to illustrate the range of remarkable molecular forms of invertebrate oxygen-binding proteins and to give some sort of classification with a view to bringing out common features. I have also attempted, where possible, to correlate structure with function, and this aspect is rapidly expanding. The first sequence and X-ray data are now appearing and subunit structures are starting to be elucidated.[4] Apart from any aesthetic attraction of the molecular forms themselves, invertebrate respiratory proteins are of great interest as a study in evolution, as examples of highly elaborate self-assembly systems, as models for co-operative phenomena, and as subjects for the study of oxygen–metal–protein interactions.[5,6]

A. NOMENCLATURE

Haemocyanins (non-haem copper) and haemerythrins (non-haem iron) are sometimes abbreviated to Hc and He by comparison with haemoglobin (Hb).

Myohaemerythrin is the monomeric haemerythrin occurring in muscle. The large invertebrate haemoglobins are frequently called erythrocruorins and those containing chlorohaem, chlorocruorins (see Table 1). I shall refer to the former as "invertebrate haemoglobins" or simply "haemoglobins": I believe this is justified inasmuch as the oxygen-binding moiety is identical in all these, i.e. protoporphyrin IX–Fe.

II. Biological Distribution

Respiratory pigments are found throughout almost all phyla of the animal kingdom, becoming increasingly indispensable in the more highly developed and evolved groups. Even some ciliated protozoans such as *Paramecium* have a "myoglobin" in the cytoplasm, but curiously chordates such as *Amphioxus* appear to have no respiratory pigment at all.[7] It is possible to make calculations of the maximal body dimensions possible in the absence of a circulatory system based on measured rates of respiration and the diffusion coefficient of oxygen. Alexander[8] suggests that for a flatworm living in aerated water the maximum possible body thickness would be about 1·0 mm, and that for a cylindrical turbellarian a diameter of up to 1·5 mm would be possible. Of course not all animals live in fully aerated water, and any circulatory system may or may not contain a respiratory pigment to increase its oxygen-transporting capacity. Nonetheless the extremely low concentrations of respiratory proteins found in some invertebrates of considerable size has led to speculation that these proteins may have a function other than oxygen transport.

Figure 1 shows the distribution of respiratory proteins in the animal kingdom in very broad outline, but there are numerous exceptions and anomalies. Nematodes may have a haemoglobin in the body cavity and a chemically distinct myoglobin in the body wall. Many annelids have a giant haemoglobin or a chlorocruorin (both of molecular weight nearly four million) dissolved in the blood, and some have instead or in addition one or more smaller, often monomeric, haemoglobins in coelomic cells.[9] One minor genus of annelids is reported to have haemerythrin, but this pigment is in fact the least widespread of all the oxygen carriers and is characteristic of three phyla: the sipunculans, the priapulids and the brachiopods.

Arthropods may have either haemocyanins or haemoglobins and it is a matter for speculation how this situation has arisen in the course of evolution. Crustaceans such as crabs and lobsters possess large haemocyanins as do *Limulus*, the horseshoe crab, scorpions, spiders and isopods. On the other hand *Daphnia* and certain shrimps (*Cyzicus, Artemia*) have moderately large haemoglobins. Only the larval forms of certain insects (e.g., *Chironomus*) have haemoglobins and these are monomeric, and haemocyanins have never been reported in insects. Echinoderms have small, intracellular haemoglobins.

4   E. J. WOOD

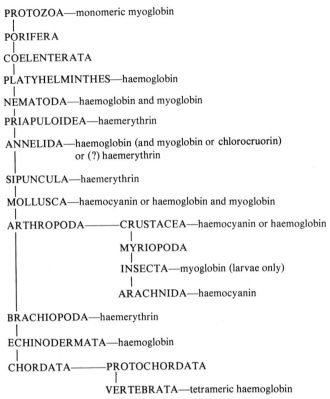

PROTOZOA—monomeric myoglobin
|
PORIFERA
|
COELENTERATA
|
PLATYHELMINTHES—haemoglobin
|
NEMATODA—haemoglobin and myoglobin
|
PRIAPULOIDEA—haemerythrin
|
ANNELIDA—haemoglobin (and myoglobin or chlorocruorin)
           or (?) haemerythrin
|
SIPUNCULA—haemerythrin
|
MOLLUSCA—haemocyanin or haemoglobin and myoglobin
|
ARTHROPODA————CRUSTACEA—haemocyanin or haemoglobin
                 |
               MYRIOPODA
                 |
               INSECTA—myoglobin (larvae only)
                 |
               ARACHNIDA—haemocyanin
|
BRACHIOPODA—haemerythrin
|
ECHINODERMATA—haemoglobin
|
CHORDATA————PROTOCHORDATA
              |
            VERTEBRATA—tetrameric haemoglobin

Fig. 1. Scheme showing the occurrence of the different types of respiratory protein in the animal kingdom.

Molluscs may possess either haemocyanins or haemoglobins, but bivalve molluscs, where they have a respiratory protein, have small (e.g., dimeric or tetrameric) intracellular haemoglobins. Gastropod haemocyanins are amongst the largest known protein molecules with molecular weights of nine million, and cephalopods (e.g., *Octopus*) have a haemocyanin of almost exactly half this size. However, some gastropods, notably the planorbid snails, have large haemoglobins instead. Again the reason for this curious divergence is not known. Typically, molluscs do not have an oxygen-binding protein in the great mass of their musculature but the exception to this is the muscles of the radula which contain a monomeric or more often a dimeric myoglobin. There does not appear to be a monomeric haemocyanin equivalent to myoglobin: gastropods have a radular myoglobin regardless of whether their circulating respiratory protein is haemoglobin or haemocyanin.[10,11] All vertebrates, with the exception of the cyclostomata, have a corpuscular, tetrameric haemoglobin and a muscular, monomeric myoglobin. Lamprey haemoglobin is monomeric and corpuscular.

Svedberg[1] noted that with the exception of the small haemoglobin of *Chironomus* larvae, all the small respiratory proteins were intracellular and all the large ones were dissolved in the blood or haemolymph. One must suspect that if high concentrations of a blood respiratory protein became essential for oxygen transport, then because of the osmotic pressure exerted and of possible loss of pigment molecules from the capillaries, including renal loss, it became necessary either to encase the protein molecules in cells or to join low molecular weight functional units together to form giant molecules which are effectively "molecular corpuscles". Thus for snail (*Helix*) haemocyanin, if the molecules were small units each carrying one molecule of oxygen, the osmotic pressure of the haemocyanin in the haemolymph would be about 2500 Pa (25 cm H_2O).[8] This is similar to the recorded pressure in the contracting ventricle of the heart of *Helix*. In fact *Helix* haemocyanin consists of aggregates of about 160 oxygen-binding units, so that the colloid osmotic pressure is reduced by a factor of 160.

III. Haemocyanins

A. THE OXYGEN-BINDING SITE

Although the structures of the protein moieties of arthropod and mollusc haemocyanins show major differences, the oxygen-binding sites, while not identical, are similar. Most of the evidence on the structure of the copper–oxygen complex has been derived by spectroscopic methods. Sequence information for haemocyanins is not available and it is not certain to which ligands the copper ions are bound.

It has long been known that the binding site of haemocyanin contains two copper ions capable of binding one molecule of oxygen. The oxy form of the protein is blue ($\lambda_{max} \sim 570$ nm) and the deoxy form colourless, and in addition only the oxy form shows the characteristic absorption band in the ultraviolet at about 340 nm. It is generally accepted that the binding site is a binuclear copper(II) peroxo complex, while the deoxy form contains copper(I). As the oxy form shows no EPR signal it is assumed that the two Cu(II) atoms are dipole coupled, and magnetic susceptibility measurements confirm that oxyhaemocyanin is diamagnetic.[12] Resonance Raman and other spectral studies reveal that oxygen is in the form of peroxide (O_2^{2-}) and that the active site probably has μ-dioxygen bridged geometry.[13,14] Calculations of copper–copper distances based on such models give values ranging from 0·35 nm to 0·50 nm. If the copper–copper distance, calculated from the EPR spectrum of nitric oxide haemocyanin, of about 0·6 nm, is the same as in deoxyhaemocyanin[15] this would imply a substantial movement of the copper ions upon oxygenation. Such a conformational change would be consistent with the oxygenation kinetics of haemocyanin.

Haemocyanin, like haemoglobin, also binds carbon monoxide but vibrational analysis shows that CO preferentially binds to one copper atom in each site with its carbon atom bound to a copper.[16] While oxygen binding is regulated by allosteric effectors, CO binding is not, and one may suppose that the protein can influence the binding of oxygen by altering the Cu–Cu distance.

The copper in haemocyanin is bound to the protein very tightly. It is not removable by chelating agents such as EDTA, and can be removed from the undenatured protein only by KCN. There is tentative evidence that histidine imidazole groups in the protein are responsible for binding the copper, and resonance Raman studies[13] indicate that the band near 340 nm in oxyhaemocyanin may be assigned to charge–transfer between N(imidazole) and Cu(II). There may in addition be another bridging ligand (Fig. 2). The 570 nm absorption band was assigned to peroxide $\rightarrow$ Cu(II) charge–transfer transitions.[14] The observed differences between the oxygen-binding sites of arthropod and mollusc haemocyanins may well be due to slight geometric alterations rather than to major structural differences in the sites.[13]

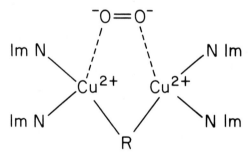

Fig. 2. Tentative structure for the oxygen-binding site of haemocyanin: the two tetragonal Cu(II) atoms are co-ordinated by histidine imidazole groups but another ligand, R, possibly phenolate, is also involved.

B. MOLLUSC HAEMOCYANINS

(1) Quaternary structure of the protein

Ultracentrifuge studies show that gastropod haemocyanins have a sedimentation coefficient of about 100 S, and those from cephalopods 50–60 S. The molecular weight of gastropod haemocyanins, calculated from accurate values for sedimentation and diffusion coefficients, is nearly nine million.[17] That for cephalopod haemocyanin is about half this value. Under the electron microscope the molecules appear to be cylinders of diameter 35 nm, and height about 35 nm in gastropod haemocyanin and about 17 nm in cephalopod haemocyanins (Fig. 3). End-on views (circles) allow ten-fold rotational

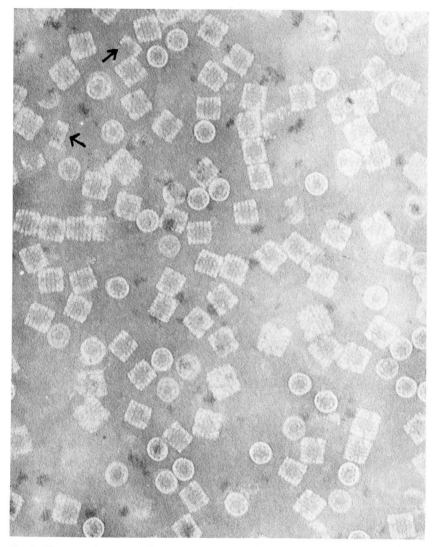

Fig. 3. Electron micrograph of negatively-stained haemocyanin from the freshwater snail, *Lymnaea stagnalis*. In end-on views of the molecule (circles) ten-fold symmetry is observed. Dimensions of the cylindrical molecule: 30 nm diam. × 32 nm height (approx.). Note the tendency of the molecules to stack: small peaks possibly corresponding to $1\frac{1}{2}$ (130 S) and double (150 S) molecules are sometimes seen in ultracentrifuge (schlieren) pictures. Occasionally one-half molecules are seen (arrowed): cephalopod haemocyanins consist almost entirely of such molecules. (Electron micrograph taken by D. Kershaw, University of Leeds.)

symmetry to be observed while side views (oblongs) suggest that the molecule is made up of a stack of six flat discs. The former is correct—the molecule does indeed have ten-fold symmetry—but the latter is artefactual. The optical image reconstruction data of Mellema & Klug[18] show that the molecule has subunits arranged helically. The fact that one sees both "back" and "front" of the molecule in side-on views, possibly due to the collapse of the cylindrical structure on the electron microscope grid, gives this appearance of six stacked plates.

Cylindrical one-half molecules (60 S) of gastropod haemocyanin will form under certain conditions and these are composed of 10 subunits arranged helically. At the end of the cylinder is an arrangement of the polypeptide chain referred to by Mellema & Klug as "collar" and this shows ten-fold symmetry. The whole molecule of gastropod haemocyanin is therefore made up of 20 subunits, arranged in sets of 10 in helical symmetry to form one-half molecules. The halves are joined in a plane perpendicular to the cylinder axis with the collar ends pointing outwards to form the native molecule with a central cavity. This model is essentially confirmed by small-angle X-ray scattering studies which also give some clues about the arrangement of the smaller subunits (see below).[19] The model produced as a result of such studies is shown in Fig. 4, and the method also gave a molecular weight of $9 \cdot 02 \times 10^6$ for the molecule.

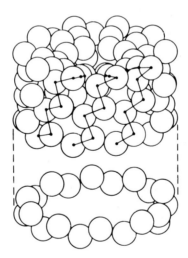

Fig. 4. Model of *Helix pomatia* β-haemocyanin constructed on the basis of low-angle X-ray scattering data. The model is built from 160 spherical subunits each of radius 2·5 nm. The domains in three one-twentieth molecules are connected by lines with points at the centres of the domains. For clarity the twenty subunits at the inner side of the bottom of the molecule are not shown. (Reproduced from ref. 19 with permission.)

(2) *The Subunits*

In addition to electron microscopic and low-angle X-ray scattering studies, extensive data on the dissociation of the native mollusc haemocyanin molecule into subunits support this model. Svedberg's early studies[1] showed that the native protein is stable only over a certain pH range (pH ~5–8). Outside this range, and depending at least partly on whether or not divalent cations such as Ca^{62+} or Mg^{2+} are present, the molecule dissociates into sub-multiples first of one-half, then one-tenth and finally one-twentieth the original size. These forms can co-exist at a given intermediate pH value (e.g., typically pH 7·9–8·5). (Cephalopod haemocyanin, of course, "starts" at the one-half stage.) These fractional molecules were originally characterized by their sedimentation coefficients and more recently by sedimentation equilibrium and high-resolution electron microscopy (Fig. 5).

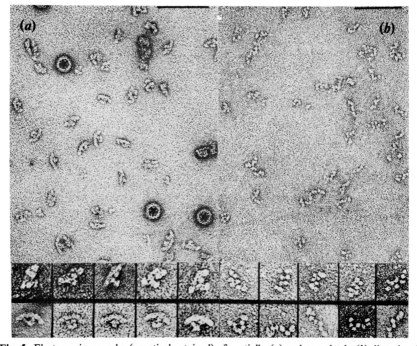

Fig. 5. Electronmicrographs (negatively-stained) of partially (*a*) and completely (*b*) dissociated *α*-haemocyanin from the snail, *Helix pomatia*. In (*a*) the pH was 7·8 and the ionic strength 0·5 and no divalent metal ion was present: some one-half molecules may be observed in both end-on and side views. The one-tenth molecules have an "arc-plus-blob" or a "parallelogram-plus-blob" appearance depending on the viewpoint. In (*b*) the pH was 10·6 and the ionic strength 0·1 when dissociation into one-twentieth molecules is practically complete. These are sometimes observed in a fairly compact arrangement, and sometimes stretched out as strings of oxygen-binding domains. Length of bar, 100 nm. Individual subunits in various orientations are picked out (below, inset). (Photographs taken from the thesis of R. J. Siezen, University of Groningen, Netherlands, and generously supplied by Professor E. F. J. van Bruggen.)

There has been much controversy over the molecular weight of the polypeptide chain(s) of gastropod haemocyanin. Early work pointed to a value of 200–250 thousand, which is much smaller than a putative one-twentieth molecule.[20] Sedimentation equilibrium studies have given values of 300 000 (in 6 M guanidium chloride)[21] and 365 000 (pH 10·5).[22] A one-twentieth molecule would in fact have a molecular weight of between 400 000 and 450 000 and values approaching this have recently been obtained by SDS-gel electrophoresis taking account of the non-linear relationship between log (mol. wt.) and mobility at high pH values and using thyroglobulin (M_r 330 000) as a calibration marker.[23] Bearing in mind that copper and protein analyses indicate that the size of the minimal functional unit (i.e. that bearing two coppers capable of binding one molecule of oxygen) has a molecular weight of about 50 000, such a moiety would contain 8–9 oxygen-binding sites. It seems likely that the fundamental polypeptide chain of gastropod haemocyanin is indeed very large and contains eight oxygen-binding domains, because (a) polypeptide chains with molecular weight of 300–400 thousand are obtained in the usual dissociating conditions (guanidinium chloride, SDS, both in the presence of mercaptoethanol), (b) it is possible to see "strings of domains" in high-resolution electron micrographs and indeed to count the domains,[22] and (c) subunits of molecular weight 50 000 may be obtained only after proteolytic digestion.[24] Limited digestion with proteolytic enzymes of different specificities has allowed individual domains to be characterized by spectroscopic, chemical and immunological methods.[25,26] The conclusion reached from such studies is that the polypeptide chain consists of a string of non-identical domains (Fig. 6).[27] This collection of linked domains presumably folds up in the appropriate way and combines with nine others to form the one-half molecule.

It is worth commenting briefly on some of the features of this interesting system. At the N-terminal end of the chain is a three-domain segment which is somewhat resistant to proteolysis and which may have a small, non-oxygen-binding portion which is non-covalently linked to the second oxygen-binding domain (Fig. 6). The so-called "collar" domains are thought to be those at the C-terminal end (g and h in Fig. 6) and the terminal domain interacts non-covalently with a neighbouring chain in the formation of a one-tenth molecule.[28] The significance of this is at present uncertain although it is noteworthy that some but not all gastropod haemocyanins have been shown to possess more than one type of polypeptide chain. Preliminary studies on the biosynthesis of gastropod haemocyanins indicate that the first products of translation of the mRNA are large and that the whole eight-domain structure is synthesized from a single stretch of mRNA.[29]

One or two collar domains per polypeptide chain may be removed from the native molecule of gastropod haemocyanin by proteolysis in the presence of Ca^{2+}. Under these conditions the "remainder" of the molecule, that is the

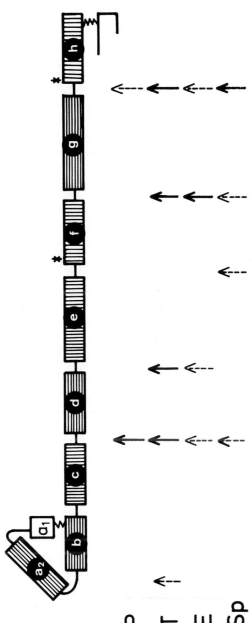

Fig. 6. Domain structure proposed for the polypeptide chain of the β_c-haemocyanin of *Helix pomatia*. The domains are represented by rectangles with a length proportional to their apparent M_r, carbohydrate included. The main cleavages are indicated by full arrows: P, plasmin; T, trypsin; E, enteropeptidase; Sp, Staphylococcal protease. Broken arrows indicate the products of limited digestion. All polypeptide stretches between the domains are susceptible to subtilisin. Non-covalent interactions between domains are represented by a zig-zag line. The isolated domains show different spectral properties indicated by horizontal or vertical shading (see refs. 26–28). (Figure kindly supplied by Professor R. Lontie, University of Leuven, Belgium.)

tubular wwall of the cylinder, undergoes extensive end-to-end polymerization to form long tubes (Fig. 7).[30] A possible function of the collar is therefore to prevent such polymerization from occurring under *in vivo* conditions.

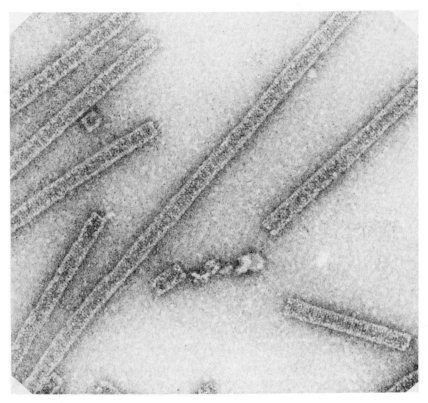

Fig. 7. Electron-micrograph of negatively-stained preparation of *Lymnaea stagnalis* haemocyanin which had been treated with trypsin in the presence of 25 mM $CaCl_2$. The collar domains have been removed and the remaining cylindrical walls of the molecules stack to form long tubes. (Electron micrograph by D. Kershaw, University of Leeds.)

(3) Primary structure

Although the amino acid compositions of a number of mollusc haemocyanins have been reported,[31] practically no sequence information is available. None of the oxygen-binding domains isolated by proteolytic digestion has so far been purified to a sufficient extent to warrant sequence studies.

Haemocyanins are glycoproteins, with carbohydrate contents varying from 1% to 8%. Again, overall compositions, but no sequences, are available (Table

2). Some mollusc haemocyanins contain unusual sugars not reported in glycoproteins from other sources, namely 3-*O*-methyl mannose and 3-*O*-methyl galactose.[32]

TABLE 2

Amino acid and carbohydrate composition of some gastropod haemocyanins expressed as mol of residue per 50 000 g protein

	Buccinum undatum	*Lymnaea stagnalis*	*Helix pomatia* (α)
Aspartic acid	51	51	46
Threonine	22	24	22
Serine	23	28	24
Glutamic acid	47	47	37
Proline	23	25	23
Glycine	25	26	26
Alanine	28	37	29
Valine	24	25	26
½ Cystine	5	7	7
Methionine	9	4	4
Isoleucine	16	20	18
Leucine	35	42	40
Tyrosine	18	14	20
Phenylalanine	25	29	22
Histidine	22	14	22
Lysine	17	30	17
Arginine	21	19	18
Tryptophan	6	7	11
N-Ac-Glucosamine	3·3	3·1	4
N-Ac-Galactosamine	0·9	2·6	2
Fucose	1·1	0·7	1·3
Xylose	0	0·5	1·6
Galactose	0·6	0·8	0·6
Glucose	0·1	0·1	0·2
Mannose	5·0	1·7	4·9
3-*O*-Me-galactose	0	present	present
3-*O*-Me-mannose	0	present	0

C. ARTHROPOD HAEMOCYANINS

(1) Quaternary structure of the protein

Ultracentrifuge studies over many years have yielded a range of sedimentation coefficients for arthropod haemocyanins. Frequently two or more types of haemocyanin molecule coexist in the haemolymph and the size of these is characteristic of the class of animal. Crustaceans and spiders typically have 16 S and 24 S components, shrimps 17 S and 39 S, scorpions 35 S and 60 S, and

Limulus, the horseshoe crab, 60 S (the values for sedimentation coefficients are approximate). The relationship between the components in some instances is very complex. Thus the haemocyanin of the shrimp, *Callianassa californiensis*, exists in the haemolymph in two forms, *C* and *I*, which can be separated.[33] The *C*-form is present as particles with a sedimentation coefficient of about 39 S, and can be reversibly dissociated into 17 S particles by removing Mg^{2+} or Ca^{2+} ions. The *I*-form is present in the haemolymph as 17 S particles and these do not associate into 39 S particles under any of the conditions tested.

The basic unit of all arthropod haemocyanins appears to be the 16–17 S component (hereafter called "16 S"), with a molecular weight of about 450 000, made up of six different polypeptide chains each of approximately the same size. The material with sedimentation coefficients of about 25 S, 39 S and 60 S results from successive dimerizations of the 16 S unit. The molecular weight of the various components of *Limulus* haemocyanin have recently been determined, by high-speed sedimentation equilibrium, by Johnson & Yphantis.[34] The values obtained for the 6 S, 24 S, 36 S and 60 S components were 69 400, 856 000, 1690 000, and 3160 000 respectively. A very simplified description of the structure of arthropod haemocyanin in comparison with that of mollusc haemocyanin is given in Table 3. In understanding the structure of arthropod haemocyanin therefore there are two questions to be considered: (a) how the 16 S hexamer successively dimerizes to give 25 S, 39 S and 60 S components, and (b) how the six 5 S polypeptide chains form the 16 S component.

TABLE 3

Comparison of the subunit structures of mollusc and arthropod haemocyanins

Mollusc haemocyanins									
No. of polypeptides[†]	1	–	2	– – – –	10	–	20	–	30 (?)
Sedimentation coef.	11 S		20 S		60 S		100 S		130 S
Mol. wt. $(\times 10^{-5})$	4·5		9·0		45		90		135
Examples			Cephalopods ⟶						
			Gastropods ⟶ – – – →						

Arthropod haemocyanins					
No. of polypeptides[‡]	1	–	6	– 12 –	24	– 48
Sedimentation coef.	5 S		16 S	25 S	36 S	60 S
Mol wt. $(\times 10^{-3})$	75		430	840	1720	3300
Examples	*Panulirus* ⟶|				
	Cancer, Homarus ⟶|				
		Scorpions ⟶|			
			Limulus ⟶|		

† Each containing eight linked oxygen-binding domains.
‡ Each consisting of a single oxygen binding domain.

(2) Polymerization of the 16 S component

Electron microscopy has given many clues as to how the 16 S unit polymerizes, although the difficulty of interpreting the pictures leaves some doubt about the exact forms of the polymers and the relationships of the subunits.[35] 16 S structure is the main component of the haemocyanin of the spiny lobster, *Panulirus*. Under the electron microscope these molecules appear as squares, rectangles or hexagons with maximum dimension

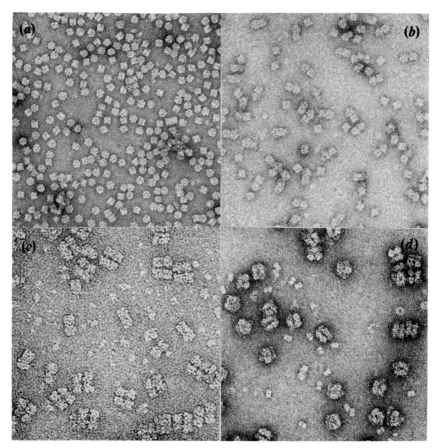

Fig. 8. Electron micrographs of negatively-stained preparations of arthropod haemocyanins. (*a*) *Panulirus interruptus* haemocyanin: the 16 S molecules present several profiles from square to hexagonal. (*b*) *Astacus leptodactylus* haemocyanin which exists mainly as dimers of the 16 S unit although some monomers may be seen. (*c*) *Hadogense bicolor* haemocyanin which exists mainly as flat plate tetramers of the 16 S unit. (*d*) *Limulus polyphemus* haemocyanin. Several profiles are observed of the 60 S molecule which is believed to be an octomer of the 16 S subunit arranged as two flat plates each composed of four 16 S subunits. All shown to the same magnification. (Photographs kindly supplied by Professor E. F. J. van Bruggen.)

10–12 nm (Fig. 8(a)): its detailed structure will be considered below. Dimerization of this type of structure to give a 25 S component occurs to produce the major component of decapod crustacean haemolymph. Under the electron microscope the structures are obviously dimeric (Fig. 8(b)) and are typically observed as a hexagonal profile connected to a square or rectangular profile. Possibly the dimer consists of two linked 16 S components, usually with rotation of one unit by 90° with respect to the other. However, in some haemocyanins patterns of two squares connected by their sides and double hexagons can also be seen.

Scorpion haemocyanins typically contain the 34 S component which usually appears in electron micrographs as large rectangles with sides approximately 23 nm (Fig. 8(c)). This molecule shows a division, as if it were a flat plate made up of two double squares connected by two narrow bridges (see Fig. 9). Finally *Limulus* haemocyanin, and also that from some spiders, contains the 60 S component. This appears to be an octomer of the 16 S structure or a dimer of the 34 S structure in which there are two layers each containing four 16 S units, possibly with some degree of rotation (Fig. 8(d)).

The role of divalent metal ions such as Mg^{2+} in the polymerization process is not completely understood. Mg^{2+} is necessary for dimer formation, whereas

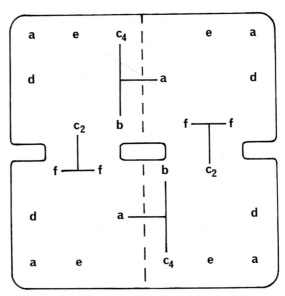

Fig. 9. Sketch to show possible but highly hypothetical arrangement of subunit polypeptide chains in a 37 S arthropod haemocyanin. Each quarter of the plate (4 × 16 S) contains six polypeptide chains and strong but noncovalent associations are shown by lines. (Redrawn after ref. 42.)

the stabilization of the tetramer seems to be essentially hydrophobic. Dissociation of all of these multimeric species (25 S, 34 S, 60 S) yields 16 S structures.

(3) Structure of the 16 S Component

The 16 S component of arthropod haemocyanins seems to be composed of six polypeptide chains each of molecular weight about 70–90 thousand and each containing one binding site for oxygen. It dissociates to 5 S components at high pH, but the resulting units are too small for any structure to be perceived by conventional electron microscopy. Schepman[36] used X-ray diffraction and electron microscopy of microcrystals to study *Panulirus* haemocyanin (16 S), and concluded that the hexamer consists of six monomers (polypeptide chains) arranged in two triangles lying in parallel planes having a common three-fold axis with a rotation of about 25°.

Studies on the polypeptide chains have indicated that there are several types. Thus it was possible to fractionate dissociated *Panulirus* haemocyanin into three components with molecular weight by SDS/gel electrophoresis of 80 000, 90 000 and 94 000.[37] Later studies showed that the 90 and 94 000 components were each composed of at least two species. Similarly Sullivan *et al.*[38] were able to separate dissociated *Limulus* haemocyanin into at least five component polypeptide chains, termed I–V, by means of chromatography on DEAE-Sephadex.

(4) Reassembly studies

The assembly of the whole (60 S) of *Limulus* haemocyanin from mixtures of different fractions has been studied.[39] Mixtures of subunits I–V or of II–V formed 60 S structures indistinguishable under the electron microscope from native ones. Mixtures lacking subunit III or subunit IV form tetramers of the 16 S unit but not octomers. All other mixtures yield either 16 S units or fail to assemble.

Similar studies were performed with spider and scorpion haemocyanin. Lamy *et al.*[40] distinguished eight subunits in the haemocyanin of the scorpion, *Androctonus*, and these are immunologically distinct. In the native molecule (34 S, tetramer of four 16 S units) there are two copies of each of four subunit polypeptide chains and four copies of the remaining four (total 24 chains). For *Eurypelma* (tarantula) haemocyanin, it is possible to distinguish seven types of subunit polypeptide chain (termed a, b, c_2, c_4, d, e, and f).[41] There are six copies of a, two each of b, c_2, and c_4, and four each of d, e, and f. A stable heterodimer may be formed *in vitro* between b and c_4. Reconstitution experiments show that at least three different subunits are required to be present for a 16 S species to form, five different for a 24 S species, six different for a 28 S species, and all seven for a stable 37 S species (as in the native molecule). On the basis of such studies it is possible to propose a tentative structure for the native molecule, showing the relationship between the subunits in the 37 S molecule (Fig. 9).[42] Presumably other spider haemocyanins are similar and *Limulus* haemocyanin is probably a dimer of this type of structure.

(5) Primary structure

So far only short amino acid sequences have been published for the polypeptide chains of arthropod haemocyanins. However, now that the subunits may be isolated and purified we may expect rapid progress in this area.[44] Sequences of up to about 30 residues from the N-terminus have been determined for *Panulirus* haemocyanin subunits.[37] The subunits of molecular weight 90 000 and 94 000 are highly homologous in this region but the 80 000 subunit is much less so. The first eleven amino acids of the two "heavy" subunits are: Asp-Ala-Leu-Gly-Thr-Gly-Asn-Ala-Glu-Lys-Glu-.

TABLE 4

Amino acid composition in residues (nearest integer) per chain of the polypeptide chains of *Eurypelma californicum* haemocyanin[43]

Chain	a	b	c_2	c_4	d	e	f	Weighted total†
Lys	46	24	33	27	33	38	29	34
His	38	22	33	21	24	26	29	27
Arg	52	39	51	23	35	40	52	43
Asp	70	68	86	73	89	94	77	81
Thr	31	36	32	30	31	33	39	33
Ser	26	37	41	40	55	33	45	40
Glu	53	60	58	77	68	63	74	65
Pro	19	31	22	32	29	28	16	24
Gly	32	45	65	84	62	54	49	54
Ala	42	30	39	56	42	38	47	42
Val	27	42	28	37	34	36	31	33
Met	9	9	13	19	12	16	17	14
Ile	25	25	17	22	28	30	18	24
Leu	48	52	47	51	63	57	68	56
Tyr	23	26	24	19	23	27	30	25
Phe	32	33	30	26	30	35	33	32
½Cys	3	9	ND	7	ND	5	ND	–
Trp	2	8	8	15	8	5	6	7
Mol. wt.	6700	6800	71 000	71 000	74 000	74 000	76 000	–

† Assuming the following percentage composition: a, 15%; b, 8·3%; c_2, 15%; c_4, 8·3%; d, 15%; e, 19%; f, 19%.

Work is also proceeding on the sequences of the chains of *Eurypelma* haemocyanin. The amino acid compositions of the seven chains show striking differences (Table 4). The *N*-terminus is proline in chains *b* and *e*, but no *N*-terminus could be found for the other chains. Chains *b* and *e* had valine as the *C*-terminal amino acid.

D. FUNCTIONAL PROPERTIES OF HAEMOCYANINS

Oxygen equilibria have been reported for many haemocyanins and all possible situations are found to occur depending upon the source of the material and the experimental conditions employed. In general, under physiological conditions, which includes the presence of Ca^{2+} or Mg^{2+} ions, haemocyanins show co-operative oxygen binding (Fig. 10). When the data are presented in the form of a Hill plot, it is found that the slope (n) at high and low oxygen tensions is 1·0, but that there is a region of higher slope at intermediate oxygen tensions. For gastropods, n typically has values of between 2 and 4, but for arthropod haemocyanins, and especially those from spiders, n may be as high as 9.[45] Within the framework of the concerted model of Monod, Wyman

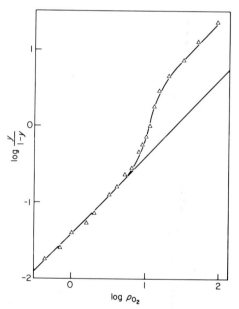

Fig. 10. Hill plot of oxygen equilibrium of *Buccinum undatum* (whelk) haemocyanin at pH 7·8 in the presence of 50 mM $CaCl_2$ and 0·5 M NaCl at 20°C. Haemocyanin concentration 1·3 mg ml^{-1}. Note that the initial and final slopes (n) are 1·0 but that there is an intermediate slope of $n > 2$.

& Changeaux, the Hill plots reflect a conformational transition from a low affinity T state to a high affinity R state. For such proteins with a large number of oxygen-binding sites the $T \rightarrow R$ transition may take place within a fairly narrow range of oxygen tensions, that is to say that addition of the first few, and of the last few, oxygen molecules has comparatively little influence on the conformation of the protein.

There are indications that it is necessary for the subunits of mollusc haemocyanin to be constrained in the cylindrical structures for co-operativity to be shown. Recent electron micrograph studies of the tubular polymers obtained from the *Helix pomatia* β-type haemocyanin (see Fig. 7) suggest a "breathing" of the molecule reflected in an increase in the cylinder diameter and a decrease in length upon oxygenation.[46] It has been known for some time that some gastropod haemocyanins under the appropriate conditions of ionic strength and pH, undergo an oxygen-linked association–dissociation process suggesting that a change of conformation is occurring.[47] The structural basis for this type of concerted change remains to be discovered.

The fundamental subunits of haemocyanin, that is the 5 S polypeptide chains of arthropod haemocyanin and the eight-domain polypeptide chains of mollusc haemocyanin, do not bond oxygen co-operatively. It is only when these are formed into polymeric aggregates that co-operative oxygen binding occurs. However, the binding by different arthropod 5 S subunits of isolated mollusc domains shows different kinetic properties.[38] The oxygen binding by whole molecules (which has long been known to be heterogeneous) presumably reflects the sum of behaviour of the different subunits or domains and presumably has a functional significance.

Many haemocyanins exhibit an alkaline Bohr effect. Oxygen binding by cephalopod haemocyanins shows a high sensitivity to pH change, and the Bohr effect is in the "normal" direction, i.e. a decrease in pH results in a decrease in oxygen affinity. In contrast many gastropod haemocyanins show a "reverse" Bohr effect at physiological pH values, that is they have the ability to combine more firmly with oxygen as the pH falls. Some gastropods, such as *Helix pomatia* have two types of haemocyanin in the blood, α and β, and on increasing the pH from 6·5 to 8·15 the oxygen equilibrium curves move in opposite directions!

Most marine gastropods show a simple reverse Bohr effect, and a number of attempts have been made to explain this curious phenomenon.[48] Recently Brix et al.[49] have made careful measurements of the oxygen-binding properties of *Buccinum undatum* haemocyanin *in vitro* and *in vivo* under normoxic and hypoxic conditions at 10°C. They found that both oxygen affinity and oxygen carrying capacity is strongly pH-dependent. This is termed a Root effect, and has been observed with the blood of some marine fishes.[50] Brix et al. suggest that these properties enable *Buccinum*, which in its natural habitat is periodically exposed to low ambient oxygen tensions and to low pH due to redox processes in sediments, to maintain a near-normal oxygen uptake down to ambient water oxygen tensions of 25–50 mm Hg.

IV. Invertebrate Haemoglobins

Haemoglobins from annelids, molluscs and arthropods were shown by Svedberg and his colleagues in the 1930s[1] to display a bewildering variety of molecular sizes, ranging from what we would call "monomeric" (i.e. molecular weight 17 000 with a single haem group), to giant aggregates with molecular weight four million. With the exception of the chlorocruorins, all of them appear to have iron–porphyrin IX as the prosthetic group just as in vertebrate haemoglobins and myoglobins. The red-green chlorocruorins have a formyl rather than a vinyl group at position 2 of the porphyrin ring: in all other respects they show a marked similarity to the annelid haemoglobins with molecular weight of 3–4×10^6.

The invertebrate haemoglobins do not appear to form a homogeneous group of proteins, and with the exception of the giant annelid haemoglobins, few of them have been studied intensively until comparatively recently. Possibly because of the paucity of detailed information, few common features can be perceived. A classification in terms of molecular and polypeptide chain size is attempted in Table 5, but no doubt our ideas will be modified in the light of future researches.

Amongst the lower phyla, haemoglobins (or myoglobins) have been reported in the protozoa, turbellaria, trematode platyhelminthes, nemertea and nematoda.[51] The haemoglobin of *Paramecium* is monomeric with a molecular weight of about 16 000 as is that from the fluke, *Dicrocoelium dendriticum.*[9] The fluke haemoglobin shows conservation of many critical residues found in both vertebrate and invertebrate haemoglobins. Nematode haemoglobin (e.g., that from *Ascaris lumbricoides*) is in contrast a multimeric molecule with a molecular weight of about 328 000 and it is proposed that it is made up of eight subunits each of molecular weight about 40 600. Unlike any of the other haemoglobins so far studied, the subunits of this haemoglobin seem to contain one haem group per 40 600 g protein rather than per 15 000-17 000 g.[9] Similarly the monomeric, body-wall myoglobin from this animal contains a single haem per 37 000 g protein and has an exceptionally high oxygen affinity with values of P_{50} of between 0·001 and 0·004 mm Hg at 20°C. It will be of great interest to discover what, if anything, these haemoglobins and myoglobins have in common with other invertebrate haemoglobins and myoglobins.

E. J. WOOD

TABLE 5

Invertebrate haemoglobins and chlorocruorins

Species	s	Mol. wt. $(\times 10^{-3})$	Minimal mol. wt.† $(\times 10^{-3})$	Chain mol. wt. $(\times 10^{-3})$	Ref.
Protozoa					
Paramecium (I)‡	1·5 S	16	16	16	9
Platyhelminthes					
Dicrocoelium dendriticum (I)	–	–	–	15·5–22·0	9
Nematodes					
Ascaris lumbricoides (E)	11·4 S	328	40	40·6	9
Ascaris lumbricoides (I)	–	37	37	37	9
Annelida					
Lumbricus terrestris (Oligochaetea) (E)	60 S	3800	17–23	12, 14, 16, 19, 31, 36	56
Eudistylia vancouveri (Sabellidae) (E)§	57·6 S	3100	26	15, 30	56
Glycera rouxii (Glyceridae) (I)	3·5 S	34·5	17	17	65
Mollusca					
Cardita borealis (Heterodont clam) (E)	–	12 000	22	290	80
Heliosoma trivolvis (Gastropoda) (E)	33·8 S	1700	18	200	56
Heliosoma trivolvis (Gastropoda) (I)	–	17·8	17·8	15	77
Anadara senilis (Arcid clam) (I)	3·1 S	25	–	13	81
(I)	4·7 S	55	–	13	81
Arthropoda					
Chironomus thummi (Insecta) (I)	2·1 S	15·4	–	15·4	82
(I)	3·1 S	34·5	–	17·2	82
Artemia salina (Anostraca) (E)	11·4 S	240	17·2	122	82
Lepidurus apus (Notostraca) (E)	19·3 S	798	17·4	34	82
Cyzicus hierosolymitanus (Notostraca) (E)	11·4 S	280	18	30	82
Holothuridae					
Cucumaria miniata (I)	2·9 S	36	–	18	92

† Per haem or per Fe.
‡ E = extracellular, I = intracellular.
§ Chlorocruorin: all the rest are haemoglobins.

A. THE OXYGEN-BINDING SITE

There is little chemical information on the environment of the haem group in invertebrate haemoglobins. In spite of great differences in the structure and function of different haemoglobins it is probable that the fundamental mechanisms underlying the combination with ligands are the same for all these pigments, including the chlorocruorins, and are closely similar to those for vertebrate haemoglobins. The only X-ray data that have appeared are those for the monomeric oxy-haemoglobin of the larvae of the insect, *Chironomus*.[52] The oxygen molecule was found to be bound to the Fe through one oxygen atom and hydrogen-bonded to a water molecule. The iron, which is $0 \cdot 17$ Å out of the plane of the haem in the deoxy form, changes its position only very slightly and moves to the proximal side upon oxygenation. The circular dichroic spectra in the Soret region of invertebrate haemoglobins show marked differences when compared with those for vertebrate haemoglobins.[53,54] The overall shape of the Cotton effect in this region is particularly sensitive to any substituent or any conformational change, but precise interpretations of these spectra must await structural information about the proteins. It therefore remains to be seen how the vastly different protein structures found in invertebrate haemoglobins and chlorocruorins modify the oxygen-binding behaviour of the haem group to produce the documented wide range of functional properties (see Section IV).

B. ANNELID HAEMOGLOBINS AND CHLOROCRUORINS

(1) Extracellular proteins

Svedberg showed that the blood of the earthworm, *Lumbricus*, as well as of other polychaete worms, contained in free solution macromolecules with sedimentation coefficients of about 60 S.[1] He obtained a molecular weight of about $2 \cdot 7 \times 10^6$ for these molecules by sedimentation equilibrium, but more recent studies have put the molecular weight nearer 4×10^6 and this has been confirmed by low-angle X-ray scattering studies.[55,56] Like the haemocyanins, these proteins are good subjects for electron microscopy, and the first pictures appeared in 1960.[57] The molecules appear to be composed of two flat hexagonal plates (Fig. 11), each being composed of six triangular subunits (or groups of subunits). There seems to be a central cavity, although in one or two of the species so far studied there may be an "extra" subunit in this space.[58] In addition a dimer of this molecule is present in some species which does not appear to be in a state of rapid dissociation–association equilibrium with the monomer.

The dimensions of the basic two-plate molecule are about 26 nm across the flats of the hexagon by 17 nm in height. Small variations on these figures have been reported and these are probably due largely to different electron microscope staining techniques. The chlorocrurorins have similar shapes and dimensions.[59]

Various models for the arrangement of subunits in the molecule have been proposed, based on interpretations of electron micrographs and on subunit data from detergent-gel electrophoresis. The latter are difficult to interpret

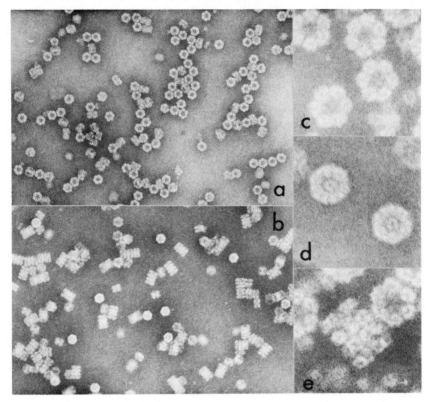

Fig. 11. Electron micrographs of negatively-stained *Euzonus micronata* haemoglobin. (*a*) Vascular fluid, (*b*) sample from leading edge of peak after passage through a column of Sepharose 4B showing dimers, (*c*) face view of monomer, (*d*) face view of dimer, (*e*) side view of dimer. Dimensions of molecule: face view, edge to edge, 25·5 nm, height, monomer, 17 mn. (Reprinted from ref. 58 with permission: electron micrography by Dr E. Schabtach, University of Oregon.)

because in many instances a number of different-sized polypeptide chains appear to be present (see Table 6). There is no doubt that the fundamental subunit or subunits is small and probably contains a single haem group. Various explanations have been proposed for the discrepant detergent-gel electrophoresis data, including suggestions that certain dimers occur which may be disulphide-linked and which are difficult to separate, and also that not necessarily all of the subunits carry a haem.

Vinogradov *et al.*[61] proposed that there may be a common type of polypeptide chain of molecular weight about 12 000, and has also presented models for *Lumbricus* haemoglobin[62] (Fig. 12). However, at present none of the models so far proposed can be accepted as definitive. The two areas of disagreement are the molecular weight per haem and the lack of consensus on the number and

TABLE 6

Subunit composition of some annelid haemoglobins [56, 60]

	Min. mol. wt. per haem	Mol wt. of chains $(\times 10^{-3})$
Oligochaetes		
Lumbricus terrestris	~23 000	12, 14, 16, 19, 31, 36
Limnodrilus gotoi	~28 000	13, 28
Achaetes		
Haemopis sanguisuga	24 800	15·5, 25·1, 51·5
Dina dubia	21 200	12, 14, 28
Polychaetes		
Eunice aphrodoitois	26 700	14·6, 20·2, 28·0, 33·1
Euzonus micronata	22 370	12·5, 14·4, 31·4, 36·7
Spirographis spallanzani (chlorocruorin)	34 000	15·8, 16·1

molecular weights of the polypeptide chains. In the case of the latter, there may well be differences between different haemoglobins, but the overall similarity of the molecules from a wide variety of species under the electron microscope suggests a basically similar subunit structure. Nonetheless such models are valuable as bases for further experimental studies and will no doubt be refined in the light of the chemical and X-ray crystallographic information now emerging.

(2) Primary Structure of Extracellular Haemoglobins

Very recently Garlick & Riggs have separated and purified some of the polypeptide chains of *Lumbricus* haemoglobin by ion-exchange chromatography in the presence of 8M urea.[63]. One of the chains (AIII in their nomenclature) was sequenced and was found to contain 157 residues giving a molecular weight, assuming one haem group, of 18 000. In agreement with this the purified chain on gel filtration in the presence of 6 M guanidinium chloride runs in the same position as sperm whale myoglobin. There are clear homologies with vertebrate haemoglobins and myoglobins. Overall there is about 15% homology with the β-chain of human haemoglobin, and both proximal and distal haem-binding histidyl residues are present. It therefore seems likely that the haemoglobin fold emerged long before vertebrates evolved.

It is interesting to note that this chain is likely to have a comparatively high (i.e. 70–80%) helical content as is found in vertebrate haemoglobins and myoglobins. This contrasts strikingly with previous estimates, from circular dichroism measurements, of about 40% α-helix for whole native *Lumbricus*

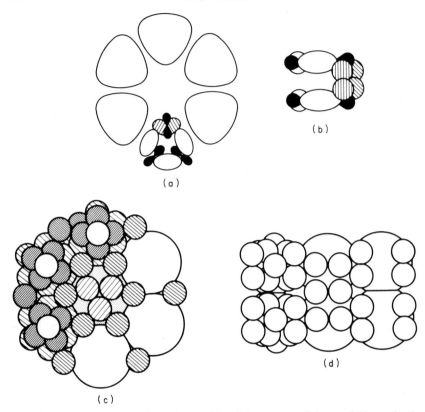

Fig. 12. Proposed structure of earthworm (*Lumbricus terrestris*) haemoglobin, and other annelid haemoglobins. (*a*), top view, (*b*), side view. Dimers of the 16 000 mol. wt. subunit are shown in solid black, the 50 000 mol. wt. subunits are white, and the 31 000 and 37 000 mol. wt. subunits are hatched. (Reproduced from ref. 62 with permission.) (*c*) and (*d*), model derived from low-angle X-ray scattering data: (*c*) shows how "central" subunits may be arranged in those annelid haemoglobins that possess central subunits. (Figures kindly supplied by Dr I. Pilz and Professor S. N. Vinogradov.)

haemoglobin.[53] It remains to be seen whether this reflects a failure of circular dichroism methods to estimate α-helical content or whether certain other chains of *Lumbricus* haemoglobin are very low in helical content. Further developments may be expected to occur with some rapidity, particularly as crystals of a suitable size for X-ray diffraction studies have recently been prepared (W. Love, personal communication).

(3) Intracellular annelid haemoglobins

In contrast to the giant extracellular annelid haemoglobins, haemoglobins with much lower molecular weight are encountered in the coelomic cells of some families of polychaetes and these are additional to the circulating extracellular proteins. Some of these are monomeric with a molecular weight of about 17 000 (one haem), whereas others are tetrameric, but in fact a variety of oligomers is commonly found. Thus that from *Glycera dibranchiata* coelomocytes had a molecular weight by osmometry of 125 000,[64] while that from *Glycera gigantea* exists as both monomers and tetramers.[65] On the basis of isoelectric focusing experiments, these haemoglobins show a high degree of heterogeneity, three to six components frequently being present. There is no information on their structures.

(4) Functional properties of annelid haemoglobins

Allen & Wyman[66] determined the oxygen equilibrium of *Arenicola* haemoglobin in 1951 and since that time many other annelid haemoglobins have been investigated by equilibrium and kinetic methods. Annelids show a multiplicity of respiratory adaptations governed both by specific environmental conditions and species-specific physiological adaptations. As is the case with haemocyanins, all possible situations occur with respect to oxygen affinity, co-operativity and Bohr effect.[67] Thus the haemoglobin of the terebellid polychaete, *Eupolymnia crescentis* has a low oxygen affinity (P_{50}, the oxygen tension at half-saturation, 36 mm Hg), no Bohr effect and no significant co-operativity ($n = 1.06$). On the other hand, *Arenicola marina* haemoglobin has a high oxygen affinity ($P_{50} = 2$ mm Hg at pH 7.4), a marked Bohr effect ($\Delta \log P_{50}/\Delta$ pH $= -0.8$ at pH 7.4), and exhibits a high degree of co-operativity ($n = 4.8$). Chlorocruorins similarly show a range of properties.[68] Such a variety of functional properties would be expected for animals whose habitats range from free-living, both fresh water and marine, to tube-dwelling. However, while it is often possible to relate the oxygen-binding properties of the haemoglobin to the life-style of the animal, we have no information which could provide a chemical explanation of how these various effects are mediated.

The smallest subunits of *Lumbricus* haemoblobin show no co-operativity, but an intermediate 10 S subunit (? corresponding to a one-twelfth molecule) obtained by David & Daniel binds oxygen co-operatively although with a higher affinity.[69] This suggests that the minimal co-operative unit is an assembly of different polypeptide chains. These findings are not supported by kinetic measurements however.[70] The Bohr effect appears to have a kinetic origin:[71] the rate constants for oxygen association were constant over the range pH 6–9, while the rate constant for oxygen dissociation increased by a factor of about 800 from pH 9 to pH 6. Kinetic studies on haemocyanins reveal a similar interplay of oxygen association and dissociation constants accounting for both positive and negative Bohr effects.[72].

Allosteric effectors of vertebrate haemoglobins, such as glycerate 2,3-bisphosphate, inositol hexaphosphate and ATP, have not been found to have any significant effect on annelid haemoglobin oxygen binding and indeed would hardly be expected to be involved in the control of oxygen binding by extracellular proteins. Interestingly, however, the oxygen affinity of some, but not all of the intracellular annelid haemoglobins is reported to be slightly but significantly depressed by ATP.[64,65] These intracellular haemoglobins typically show high oxygen affinities (P_{50} = 5–8 mm Hg) and a small degree of co-operativity ($n = 1\cdot0-1\cdot5$).

C. MOLLUSC HAEMOGLOBINS

Although haemocyanins are fairly typical of gastropods and cephalopods, haemoglobins are nevertheless found in some molluscs. Some bivalves have small but oligomeric intracellular haemoglobins while some gastropods, notably the planorbid snails, have large extracellular haemoglobins. Many gastropods, as well as opisthobranchs, have a monomeric or dimeric radular muscle myoglobin in addition to a circulating haemoglobin or myoglobin. The data so far obtained with a range of molluscs are summarized in Table 5.

(1) Extracellular gastropod haemoglobins

Planorbid snails (*Planorbis, Heliosoma*, and *Biomphalaria* spp.) characteristically have large extracellular haemoglobins in the blood with sedimentation coefficients of approximately 33 S[1] and molecular weight of 1 700 000.[73,74] These molecules may be studied under the electron microscope when they are observed as circular structures about 20 nm in diameter. Typically many "broken" structures are also seen, and so far it has not been possible to obtain an impression of the side view of the molecule (Fig. 13). The circular forms show a ten-fold symmetry, as is found in mollusc haemocyanins, but at present no other structural information can be obtained from the pictures.

Denaturation of the molecules in 6 M guanidinium chloride or in SDS, in the presence of 2-mercaptoethanol, followed by molecular weight determination give values of about 220 000 for the polypeptide chain. As the minimal molecular weight per haem is probably 17–18 thousand, it seems likely that each of these chains contains 8–12 oxygen-binding domains, and that 10 of these chains assemble in a ring to form the native molecule. Limited proteolysis yields a selection of sub multiples, which are still capable of binding oxygen, and with molecular weight of 18 000, 40 000, 60 000, and higher.[54,74] Thus there are remarkable parallels with molluscan haemocyanins which also have multi-domain polypeptide chains.

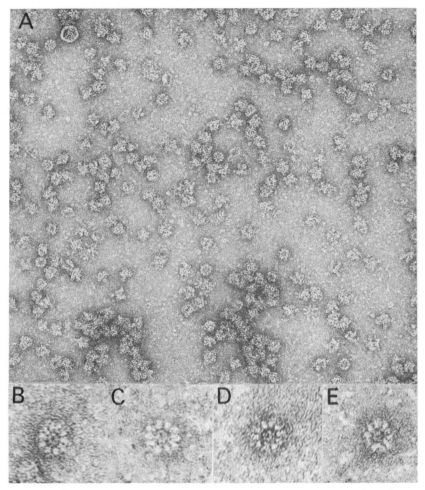

Fig. 13. Electron micrographs of negatively-stained haemoglobin from the snail, *Heliosoma trivolvis*. In (A) some of the ten-membered ring structures may be seen, and (B), (C), (D), and (E) show higher magnifications of such structures, approx. diameter 20 nm. (Electron micrographs from ref. 74 kindly suppled by Dr N. C. Terwilliger, Dr R. C. Terwilliger and Dr E. Schabtach, University of Oregon.)

Gastropod haemoglobins typically bind oxygen co-operatively at least at some pH values, and values of n between 1·7 and 3·4 have been reported.[73] On the other hand, the haemoglobin from *Biomphalaria glabrata* was reported to have a Hill coefficient of only slightly greater than 1·0 at pH 7·0.[75] Oxygen affinities of planorbid snail haemoglobins are relatively high ($P_{50} = 2$–9 mm Hg) but only *Heliosoma* haemoglobin has so far been shown to have a Bohr effect.[75] The domains isolated by proteolysis had relatively high affinities and lacked both Bohr effect and co-operativity.

(2) Myoglobins from the radular muscle of gastropods

Myoglobins, which are either monomeric (mol. wt. 17 000) or more frequently dimeric are found in the radular muscles and also in the stomach muscles and the nervous system of many gastropods.[76] Many of these proteins have been purified and their amino acid compositions and some sequences have been reported. Thus the myoglobins of *Heliosoma* are very similar in amino acid composition to the subtilisin-generated subunit of the vascular haemoglobin.[77] The sequence of *Busycon canaliculatum* myoglobin (dimeric) is homologous with, but about 80% different from, all other related globins of known sequence (Fig. 14). It is, in fact, more similar in sequence to human haemoglobin β-chain than to the myoglobin of another gastropod, *Aplysia limacina* (75% and 94% difference, respectively). The interdimer interactions occur in different regions or involve different amino acids compared with vertebrate haemoglobins, because no obvious similarities are discernible with residues known to be present in the contact regions ($\alpha_1\beta_1$ or $\alpha_1\beta_2$) of other haemoglobins.

Busycon myoglobin shows slight co-operative oxygen binding ($n \sim 1\cdot1$), and that from another gastropod, *Nassa mutabilis*, has $n = 1\cdot5$.[79] No myoglobin shows an appreciable Bohr effect. The oxygen affinity (P_{50}) of *Nassa* myoglobin dimer over the range pH $6\cdot5$–$8\cdot0$ is $4\cdot7$ mm Hg and calculations show that the oxygen affinity for the second oxygen molecule increases about $2\cdot2$-fold. It will be interesting to discover which amino acids in these dimeric myoglobins permit the protein–protein interactions necessary to stabilize the dimer, and also how co-operativity is mediated.

(3) Haemoglobins from bivalve molluscs

Bivalve molluscs may have either extracellular or intracellular haemoglobins, but few of these are well characterized and no doubt new structures await discovery. In addition, many species do not have a respiratory pigment.

The heterodont clam, *Cardita borealis*, has an extracellular haemoglobin, of molecular weight 12×10^6, which is seen in the electron microscope as a long rod-shaped polymer of width 21–36 nm and length 36–120 nm.[80] Surprisingly for such a large multi-subunit structure, the pigment does not exhibit a Bohr effect and does not bind oxygen co-operatively. Interestingly, dissociation and digestion studies suggest that this haemoglobin has a very large subunit

NA1 NA2 A1 A2 A3 A4 A5 A6 A7 A8 A9 A10 A11 A12 A14 A15 A16 AB1 AB2 AB3 AB4 AB5 B1 B2 B3
Gly-Leu-Asp-Gly-Ala-Gln-Lys-Thr-Ala-Leu-Lys-Glu-Ser-Trp-Lys-Val-Leu-Gly-Ala-Asp-Gly-Pro-Thr-Met-Met-

B4 B5 B6 B7 B8 B9 B10 B11 B12 B13 B14 B15 B16 C1 C2 C3 C4 C5 C6 C7 CD1 CD2 CD3 CD4 CD5
Lys-Asn-Gly-Ser-Leu-Phe-Gly-Leu-Phe-Lys-Thr-Tyr-Pro-Asp-Thr-Lys-His-Phe-Lys-His-Phe-Asp-

CD6 CD7 CD8 D2 D3 D4 D5 D6 D7 E1 E2 E3 E4 E5 E6 E7 E8 E9 E10 E11 E12 E13 E14 E15 E16
Asp-Ala-Thr-Phe-Ala-Ala-Met-Asp-Thr-Gly-Val-Gly-Lys-Ala-His-Gly-Val-Ala-Val-Phe-Ser-Gly-Leu-Gly-

E17 E18 E19 E20 EF1 EF2 EF3 EF4 EF5 EF6 EF7 EF8 F1 F2 F3 F4 F5 F6 F7 F8 F9 FG1
Ser-Met-Ile-Cys-Ser-Ile-Asp-Asp-Asp-Cys-Val-Asx-Gly-Leu-Ala-Lys-Lys-Leu-Ser-Arg-Asn-His-Leu-Ala-

FG3 FG4 FG5 G1 G2 G3 G4 G5 G6 G7 G8 G10 G11 G12 G13 G14 G15 G16 G17 GH3 GH4 GH5 H1 H2 H3
Arg-Gly-Val-Ser-Ala-Ala-Asp-Phe-Lys-Leu-Glu-Ala-Val-Leu-Glu-Lys-Leu-Ala-Val-Phe-Lys(Glx-Phe-Leu-Asp-Glu-Ala-Thr-Gln-Arg)-

H4 H5 H6 H7 H8 H9 H10 H11 H12 H13 H14 H15 H16 H17 H18 H19 H20 H21 H22 H23 H24 H25
Lys-Ala-Thr-Asp-Ala-Gln-Lys-Asp-Ala-Asp-Gly-Ala-Leu-Leu-Thr-Met-Leu-Ile-Lys-Ala-His-Val

Fig. 14. Amino acid sequence of radular myoblogin from the whelk, *Busycon canaliculatum*.[78]

polypeptide chain (mol. wt. 300 000), containing a number of oxygen-binding domains (mol. wt. 17 000) which may be separated by gentle proteolysis. In this respect it is similar to the extracellular haemoglobin of *Planorbis*.

In contrast, clams of the primitive family Arcidae have both dimeric and tetrameric, intracellular haemoglobins. The dimeric and tetrameric haemoglobins are distinct and have molecular weight of approx. 34 000 and 68 000 respectively. Those from the blood clam, *Anadara senilis*, are easily separable by ion-exchange chromatography and have distinct oxygen-binding properties as well as distinct amino acid compositions.[81] The dimer binds oxygen with a higher affinity than does the tetramer but both show co-operativity ($n =$ 1·5–2·0 and 1·6–2·3, respectively). The functional significance of the possession of two types of haemoglobin molecules remains to be determined.

D. ARTHROPOD HAEMOGLOBINS

The occurrence of haemoglobins in arthropods is described as "sporadic", and is confined to the branchiopods and a few insects. On the basis of the little information available, it is possible to divide these haemoglobins into classes according to source and molecular properties as follows: those from insects, those from branchiopods lacking a carapace and those from branchiopods which have a carapace.[82]

(1) Insect haemoglobins

Insect larvae haemoglobins are the only ones that are well characterized, and X-ray crystallographic studies on the structure of the monomeric haemoglobin from *Chironomus* have already been mentioned (p. 23). However, Keilin & Wang[83] mention haemoglobins occurring in Hemiptera, e.g. *Macrocrixa geoffroyi*, the large water boatman, and report oxygen-binding studies on the haemoglobin of the larvae of the horse botfly, *Gastrophilus*.

Insect larvae haemoglobins are monomeric or dimeric and bind oxygen with a fairly high affinity (e.g., $P_{50} = 5·6$ mm Hg); as they are intracellular in tissues rather than in blood they perhaps would be better described as myoglobins. They exhibit polymorphism, and several of the *Chironomus* haemoglobins have been sequenced.[84]

(2) Haemoglobins of carapaceless arthropods

It has been known for some time that certain branchiopods of the class Anostraca can possess extracellular haemoglobins and the best-studied of these is that from the brine shrimp, *Artemia salina*. This animal, which inhabits salt lakes that show wide fluctuations in oxygen concentrations due to changes in salinity and temperature, has a remarkable adaptive behaviour. Not only can it adapt by gaining or losing haemoglobin,[85] but also it possesses three types of haemoglobin (I, II, III) which are adapted differently to varied environmental factors. Thus the oxygen affinities of these are Hb-III > Hb-II > Hb-I, and hypoxic conditions increase the production of Hb-III (Table 7).[86] The genetics of *Artemia* haemoglobins have also been studied and Bowen *et al.*[87] found that the three haemoglobins could give rise to ten situations: no detectable haemoglobin, all three, any one, or any combination of two haemoglobins.

All three haemoglobins of *Artemia* have similar sedimentation coefficients (11 S) corresponding to a molecular weight of about 240 000, and it is reported that a different haemoglobin, of similar size, is synthesized in the hatching nauplius (Hb-N) and this is not present in the adult.[88] Each of the haemoglobins seems to be composed of two identical polypeptides of molecular weight 110–130 thousand: earlier reports of polypeptide chains with molecular weight of 13–17 thousand may have resulted from proteolytic action.[82] These large molecular weights suggest that the subunits are composed of about seven linked oxygen-binding domains.

TABLE 7

Proportions of the three haemoglobins present in *Artemia* haemolymph following acclimation to different oxygen tensions[86]

Acclimation (% saturation with oxygen)	Hb-I (%)	Hb-II (%)	Hb-III (%)
100	25·7	73·7	–
50	22·7	76·8	–
30	22·2	77·2	+
20	9·1	77·6	12·9
10	5·6	70·3	24·0

+, Present in small amounts.

(3) Haemoglobins from carapaced arthropods

In contrast to the haemoglobins of the carapaceless arthropods, those of the carapaced ones (Notostraca, Cladocera, Conchostraca), so far studied, have high molecular weights in the native form but seem to be composed of low molecular weight subunit polypeptide chains. Each of the latter is believed to be a unit of molecular weight of about 34 000 carrying two haem groups. The number of polypeptide chains per native molecule varies with the class of animal and is believed to be 24 in *Lepidurus*, the tadpole shrimp (mol. wt. 798 000),[89] 20 in *Daphnia* (mol. wt. 420–670 thousand), and 10 in *Cyzicus*, the clam shrimp (mol. wt. 220–280 thousand).[82] The molecular weight of the polypeptide chains from each of these as determined by SDS-gel electrophoresis are not quite the same.

Few oxygen-binding studies on these haemoglobins have so far been reported, although it is known that *Cyzicus* haemoglobin binds co-operatively ($n = 2 \cdot 3$), and is reported to have an exceptionally high affinity for oxygen ($P_{50} = 0 \cdot 035$ mm Hg at pH $7 \cdot 2$ and $28 \degree C$). This is suggested to be an adaptation to conditions of extreme hypoxia, with haemoglobin as an oxygen store.[90] Five-fold symmetry was observed in electron micrographs of *Cyzicus* haemoglobin.[91]

E. HOLOTHURIAN HAEMOGLOBINS

The intracellular haemoglobin of the sea cucumber (*Cucumaria miniata*) has been studied and has a molecular weight in the oxy form of about 36 000 and a subunit molecular weight of 18 000.[92] The oxy form appears to dimerize upon deoxygenation and shows some degree of co-operativity in its oxygen binding ($n = 1 \cdot 8$ at pH $7 \cdot 4$). The N-termini of a number of holothurian haemoglobins is blocked.[93] It will be interesting to compare the structures of these haemoglobins with those of the vertebrates to which this group of animals is closely related.

V. Haemerythrins

The haemerythrins comprise the third group of oxygen-binding respiratory pigments found in nature.[94,95] In many ways they are the simplest of the respiratory proteins and are found only in a limited number of relatively minor phyla. Even when they occur as oligomeric aggregates, the subunits are practically identical and there appears to be little or no co-operativity in oxygen binding. The possession of a relatively low molecular weight subunit (13 500) containing two iron atoms and the ease with which the protein crystallizes has meant that haemerythrin is much more tractable for study than either haemocyanins or invertebrate haemoglobins. Curiously the B2 subunit of the enzyme nucleoside diphosphate reductase is similar in structure to haemerythrin.[96]

Haemerythrin occurs as an octomer of molecular weight 108 000 in cells in the coelomic fluid of sipunculids, priapulids, and brachiopods. Each of the eight subunits contains two Fe atoms (but no haem) each of which is capable of binding one molecule of oxygen reversibly. Not all haemerythrins are octomeric however. A trimeric form has been found in the sipunculid *Phascolosoma lurco*,[97] and monomeric myohaemerythrin occurs in the retractor muscles of another sipunculid, *Thermiste pyroides*,[98] which animal also has an octomeric haemerythrin in its erythrocytes. This situation is reminiscent of myoglobin and haemoglobin in vertebrates.

A. PRIMARY STRUCTURE

Complete amino acid sequences have been determined for several haemerythrins and a myohaemerythrin[99] and these are shown in Fig. 15. Invariant sequences are enclosed in boxes and although the degree of sequence identity between coelomic haemerythrins and myohaemerythrins is just over 40%, the tertiary structures by X-ray crystallography seem to be very similar (see below). There are presumably differences in surface amino acids which cause the haemerythrins to form octomers (or trimers) while the myohaemerythrins remain monomeric. This situation is similar to that of vertebrate myoglobins and haemoglobins.

The two coelomic haemerythrin sequences shown in Fig. 15 show a much greater degree of sequence homology than each does with myohaemerythrin, especially between residues 19 and 66. An interesting feature is the possession of 2 to 4 proline residues in the first 12 N-terminal residues. One would therefore expect a non-helical conformation in this region. Another interesting feature is a cysteine residue at position 50 in haemerythrins which is absent in myohaemerythrin. If this cysteine is blocked the octomer dissociates. The trimeric haemerythrins, like myohaemerythrins, do not have a cysteine at position 50.[94]

Haemerythrin variants have been found in the erythrocytes from a single species. Thus five variants of *Phasocolopsis gouldii* haemerythrin have been identified and these are of two major classes, A and B.[100] Four variants of haemerythrin A arise from substitution of Thr for Gly at position 79 and of Ala for Ser at position 96 (in Fig. 15 the sequence of the major component of *P. gouldii* haemerythrin is the one given). The B variant, of which there appears to be only one type has Gly at 79 and Ala at 96, and Glu, Asp and His at positions 63, 78 and 92 instead of Gln, Glu and His respectively. The function of these variants is unknown.

B. TERTIARY AND QUATERNARY STRUCTURE

The major structural feature of both haemerythrin and myohaemerythrin chains is the possession of four approximately parallel sections of α-helix which together account for roughly 70% of the sequence.[100] This is true despite the differences in sequence between haemerythrins and myohaemerythrins. In addition to these four major helical portions (A—D) it seems likely that the short carboxy-terminal stub is also helical (E). The arrangement of these helices can be seen in Fig. 16 and there is obviously little or no β-structure. It seems

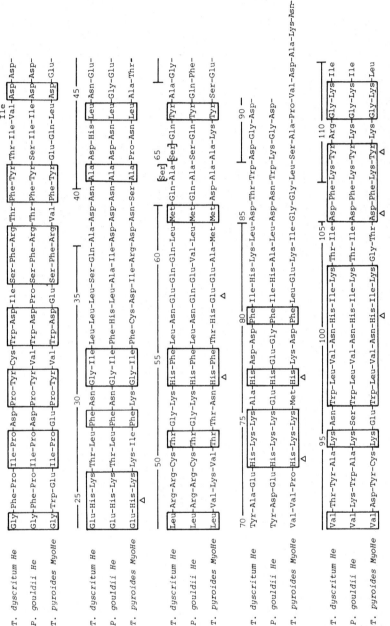

Fig. 15. Amino acid sequence of *T. dyscritum* haemerythrin in comparison with the sequences for *P. gouldii* haemerythrin (major component) and *T. pyroides* myohaemerythrin. Invariant residues are enclosed in boxes. Residues implicated as iron ligands in *T. dyscritum* haemerythrin are indicated by △. The continuous lines show the helical regions, A–E, in haemerythrin.[99]

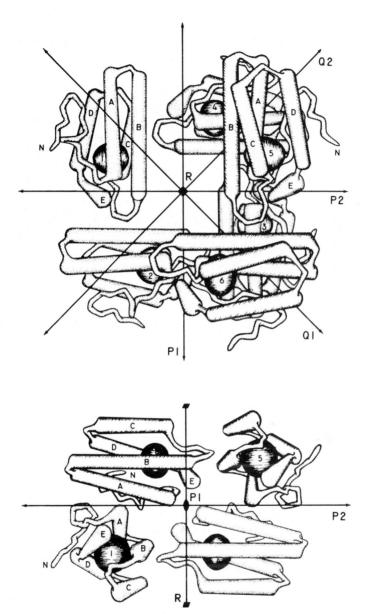

Fig. 16. Quaternary structure of haemerythrin octomer drawn from computer-generated schematics. The dyads P1 and P2 and the tetrad R are parallel to the x, y and z co-ordinates: subunits are drawn as schematic representations of the polypeptide chain. The N-terminal arm is designated by N and helical segments are labelled A–E. The dimeric iron centres are represented by oblate spheroids numbered according to the subunit to which they belong. (Taken from ref. 100 with permission.)

likely that the five extra amino acid residues in myohaemerythrin (Fig. 15) can be accommodated by the formation of a small loop between haemerythrin position 90 and 91. In the model this would represent an insertion into the CD corner.

A good deal is known about the location of the iron atoms. Each subunit, or the whole molecule in the case of myohaemerythrin, holds two iron atoms and their positions are reasonably precisely known from native anomalous scattering data at a resolution of 0·28 nm. The iron atoms are close together and are shown in Fig. 16 as oblate spheroids. Given this arrangement only a limited number of amino acid side-chains are capable of participating in metal co-ordination, and many of these can be ruled out because they are apolar or have very high or very low pK values and are therefore unlikely to form bonds with metal cations. On the basis of this type of evidence, along with an improved analysis of X-ray data for methaemerythrins at 0·28 nm, it is proposed that the iron ligands are His-25, His-54, Glu-58, His-73, His-77, His-101, Asp-106, and Tyr-109.[101] These proposed ligand positions fit well with the suggested mode of binding, that is, in a face-sharing bioctahedron. Furthermore all eight positions are conserved in the sequences so far determined.

What can be said about the arrangement of subunits in the octomer? As the subunits are identical or nearly so, each must be in an equivalent or quasi-equivalent environment. This restricts the number of possible geometric arrangements. In fact X-ray studies show that the octomer is composed of two layers, each being a square of four subunits arranged in an end-to-side fashion.[102] This arrangement is illustrated in Fig. 16 and has been described as resembling a square doughnut. It seems that the two helices in the region between residues 19 and 66 provide major points of subunit contact for octomer formation. Furthermore, it seems reasonably certain that haemerythrin monomers are essentially isostructural with myohaemerythrin. These factors, taken together with the full conservation of the putative iron ligands, allow inferences to be made about subunit contacts in octomeric haemerythrins. In fact in the proposed model the subunits interact via three types of subunit interface, within layers, between subunits of different layers, and across the P axes (see Fig. 16) i.e. between the C-terminus of the E stub and the N-terminus and A helix of one subunit with the same area of another such as between subunits 2 and 6 in the figure. It seems likely that the strongest interactions are between subunits related by one of the two-fold axes, that is those through the corners of the octomer, where side-chains are close enough to interact. Examples are Asp-23 ($-CO_2H$), and Thr-27 ($-OH$) and Arg-49 (guanidinium), although this latter is also close to Thr-19. Coulombic attraction between Arg-15 and Asp-42 is possible: the terminal oxygen of Asn-30 could be hydrogen-bonded to the backbone amide of Trp-10 or Lys-26

(E–N). Fewer interactions occur between four-fold related subunits: for example the BC corner and E stub at the end of one subunit are in contact with parts of the B and C helices at the side of a neighbouring subunit (cf. contacts between subunit 1 and 2 in Fig. 16).

C. THE OXYGEN-BINDING SITE

The spectrum of haemerythrin and some of its derivatives is shown in Fig. 17, and spectral studies have played a large part in the elucidation of the structure of the iron–oxygen complex. Model oxo-bridged complexes of the type LnFe(III)-O-Fe(III)Ln for example have very similar spectra to those of methaemerythrin and oxyhaemerythrin.

There seems no doubt that the two iron atoms are close together probably with the O_2 molecule bridging them. X-ray studies give the distance between iron centres as 0·34 nm, and putative ligands holding them in the protein matrix have been mentioned above. In fact the iron centre exists in a cage insulated from the surrounding medium by the four main helices on the sides, and by the E-stub and BC corner at one end, and by the AB and CD corners at the other end. Based on steric considerations the binding sites for O_2 are probably in the cavity opposite the apex of tyrosine ligands.

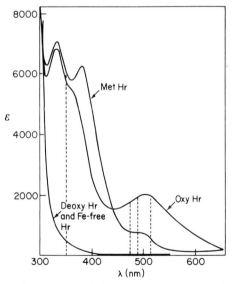

Fig. 17. Visible absorption spectra of haemerythrin and derivatives. That shown as "methaemerythrin" is with ligands such as Cl^-, OH^-, and OCN^-: other anionic ligands such as N_3^- and SCN^- have much stronger absorptions near 500 nm. (Taken from ref. 94 with permission.)

Despite the large amount of structural information, the physiological function of haemerythrin is almost entirely unknown. The oxygen affinity is relatively high and oxygen appears to be bound non-co-operatively even in the octomer. Values of no greater than 1·1 have been recorded for the Hill coefficient, n, and kinetic measurements indicate that the subunits in the octomer behave independently.[103] There is no Bohr effect, but perchlorate acts as an allosteric effector and although this is not expected to have any physiological significance it is possible that perchlorate is mimicking some other anion which does have physiological importance. Mangum & Kondon[104] have studied some of the properties of intact coelomocytes. These are large (4–25 μm diameter) biconvex discs, and it has been estimated that 1 μl of coelomic fluid contains 9×10^4 cells, each containing approximately 2 ng haemerythrin or 10^{10} molecules. The oxygen binding affinity of intact coelomic cells was lower than that of haemerythrin solution ($P_{50} = 6\cdot6$ mm Hg for cells, cf. 3·07 for solution at pH 7·4 and 22°C).

VI. Biosynthesis of Invertebrate Respiratory Proteins

There has been very little work on the biosynthesis of the invertebrate respiratory proteins described. It would, for example, be very interesting to know whether the multidomain structures frequently encountered are synthesized as a single polypeptide chain. However, in most cases even the site of biosynthesis is not known with any degree of certainty. Wells & Dales[68] mention that the biosynthesis of chlorocruorohaem in the heart-body of *Flabelligera affinis*, and the extravasal tissue of *Sabella penicillus*, resembles that in the haemoglobin-containing polychaetes. Several workers have investigated the copper content of various organs of haemocyanin-containing animals and have suggested possible sites of haemocyanin biosynthesis. Thus the blood-gland of opisthobranch species (Doridoidea) has the highest copper content of all the organs and was suggested as a possible site of haemocyanin synthesis.[105] From amino acid incorporation experiments it was concluded that the branchial gland of *Octopus* was the site of haemocyanin synthesis in these animals.[106] It was reported that the cells of this latter gland possess vacuoles containing granules that look like haemocyanin. Similarly the branchial gland of the cuttlefish, *Sepia*, has a high copper content and is considered to have a possible role in haemocyanin synthesis.[107]

Although the hepatopancreas of gastropods contains the highest levels of copper of all the organs, Sminia[108] has proposed that "pore cells" occurring in the connective tissue of snails is a more likely site of haemocyanin synthesis. The basis for this suggestion is that these cells contain crystalline arrays which may well be haemocyanin crystals. Similar crystals of haemoglobin are observed in the pore cells of *Planorbis*,[109] while blood cells ("cyanoblasts") apparently containing plates of haemocyanin crystals were reported in *Limulus*.[110]

Messenger RNA has recently been isolated from the connective tissue of *Lymnaea* and *Planorbis* and after translation in one *in vitro* system, products were identified immunologically as haemocyanin or haemoglobin, having similar molecular weight to native haemocyanins and haemoglobins on SDS-gels.[29] These preliminary findings suggest that the multi-domain polypeptide chains are synthesized from a single stretch of mRNA. These findings are in contrast to those of Moens *et al.*[114] who translated mRNA from *Artemia* (which is also believed to have a multi-domain polypeptide chain) and found only material of molecular weight approximately 17 000 which could be identified as haemoglobin. However, this synthesis took place on large (200–280 S) polysomes. No doubt interesting developments in this area can soon be expected and it is clear that synthetic studies, including determination of the size of mRNA molecules, can make valuable contributions to our knowledge of protein structure.

VII. General Concluding Remarks

The invertebrate respiratory pigments represent an attractive and interesting group of proteins with which to work, but their very diversity frustrates the collection of information and the collation of general or common features. However, patterns are beginning to emerge and the next few years should see significant developments in our knowledge and understanding. Not only will the complex architecture and self-assembly processes be elucidated but also studies on their biosynthesis will reveal many interesting facets of the way in which the molecules are constructed. Not least it is hoped that in addition, the investigation of the biological function of these proteins will give us an insight into the way in which the properties of a very few types of oxygen-binding site can be modified to cope with an enormous variety of life-styles.

REFERENCES

1. Svedberg, T. & Pedersen, K. O. (1940). *The Ultracentrifuge*. Clarendon Press, Oxford.
2. Carlson, R. M. K. (1975). NMR spectrum of living tunicate blood cells and the structure of the native vanadium chromogen. *Proc. Natl. Acad. Sci U.S.A.* **72**, 2217–2221.
3. Macara, I. G., McLeod, G. C. & Kustin, K. (1979). Vanadium in tunicates: oxygen-binding studies. *Comp. Biochem. Physiol.* **62A**, 821–826.
4. Wood, E. J. (1979) Invertebrate Respiratory Proteins. *Nature* (*London*) **281**, 341–342.
5. Hendrickson, W. A. (1977). The molecular architecture of oxygen-carrying proteins. *Trends Biochem. Sci.* **2**, 108–111.

6. Antonini, E. & Brunori, M. (1974). Transport of oxygen: respiratory proteins. In *Molecular oxygen in biology: topics in molecular oxygen research* (Hayaishi, O., ed.), pp. 219–274. North-Holland, Amsterdam.

7. Lemberg, R. & Legge, J. W. (1949). *Haematin Compounds and Bile Pigments*, pp. 305–335. Interscience, New York and London.

8. Alexander, R. McN. (1979). *The Invertebrates*, pp. 182–185. Cambridge University Press, London.

9. Terwilliger, R. C. (1980). Structures of invertebrate hemoglobins. *Am. Zool.* in the press.

10. Berthier, J. (1947). Localisation de l'erythrocruorine chez les Planorbes et les Limnées. *C. R. Acad. Sci.* **225**, 957–959.

11. Bannister, W. H., Bannister, J. V. & Micallef, H. (1968). Occurrence of hemoglobin in the radular muscles of some prosobranch molluscs. *Comp. Biochem. Physiol.* **24**, 1061–1063.

12. Dooley, D. M., Scott, R. A., Ellinghaus, J., Solomon, E. J. & Gray, H. B. (1978). Magnetic susceptibility studies of laccase and oxyhaemocyanin. *Proc. Natl. Acad. Sci. U.S.A.* **75**, 3019–3022.

13. Friedman, T. B., Loehr, J. S. & Loehr, T. M. (1976). A resonance Raman study of the copper protein, hemocyanin. New evidence for the structure of the oxygen-binding site. *J. Am. Chem. Soc.* **98**, 2809–2815.

14. Eickman, N. C., Himmelwright, R. S. & Solomon, E. I. (1979). Geometric and electronic structure of oxyhemocyanin: spectral and chemical correlations to met apo, half met, met, and dimer active sites. *Proc. Natl. Acad. Sci. U.S.A.* **76**, 2094–2098.

15. Schoot-Uiterkamp, A. J. M., van der Deen, H., Berendsen, H. J. C., & Boas, J. F. (1974) Computer simulation of the EPR spectra of mononuclear and dipolar coupled Cu(II) ions in nitric oxide and nitrite-treated hemocyanins and tyrosinase. *Biochim. Biophys. Acta* **372**, 407–425.

16. Van der Deen, H. & Hoving, H. (1979). An infrared study of carbon monoxide complexes of haemocyanins. Evidence for the structure of the CO-binding site from vibrational anlaysis. *Biophys. Chem.* **9**, 169–179.

17. Wood, E. J., Bannister, W. H., Oliver, C. J., Lontie, R. & Witters, R. (1971). Diffusion coefficients, sedimentation coefficients and molecular weights of some gastropod haemocyanins. *Comp. Biochem. Physiol.* **40B**, 19–24.

18. Mellema, J. E. & Klug, A. (1972). Quaternary structure of gastropod haemocyanin. *Nature (London)* **239**, 146–150.

19. Berger, J., Pilz, I., Witters, R. & Lontie, R. (1977). Studies by small-angle X-ray scattering of the quaternary structure of the β-haemocyanin of *Helix pomatia*. *Eur. J. Biochem.* **80**, 79–82.

20. Brouwer, M. & Kuiper, H. (1973). Molecular weight analysis of *Helix pomatia* α-haemocyanin in guanidine hydrochloride, urea and sodium dodecylsulphate. *Eur. J. Biochem.* **35**, 428–435.

21. Quitter, S., Watts, L. A., Crosby, C. & Roxby, R. (1978). Molecular weight of aggregation states of *Busycon* hemocyanin. *J. Biol. Chem.* **253**, 525–530.

22. Siezen, R. J. & van Bruggen, E. F. J. (1974). Structure and properties of hemocyanin. *J. Mol. Biol.* **90**, 77–89.

23. Gielens, C., Verschueren, L. J., Préaux, G. & Lontie, R. (1979). Fragmentation of crystalline β-haemocyanin of *Helix pomatia* with plasmin and trypsin. Location of the fragments in the polypeptide chain. *Eur. J. Biochem.* **103**, 463–470.

24. Lontie, R., De Ley, M., Robberecht, H. & Witters, R. (1973). Isolation of small functional subunit of *Helix pomatia* haemocyanin after subtilisin treatment. *Nature (London)* **242**, 180–182.

25. Gullick, W. J., Herries, D. G. & Wood, E. J. (1979). Characterization of domains obtained from a mollusc haemocyanin by limited proteolytic digestion. *Biochem. J.* **179**, 593–602.

26. Gielens, C., Préaux, G. & Lontie, R. (1977). Structural investigations on β-haemocyanin of *Helix pomatia* by limited proteolysis. In *Structure and Function of Haemocyanin* (Bannister, J. V., ed.) pp. 85–94. Springer-Verlag, Berlin.

27. Gielens, C., Verschueren, L. J., Préaux, G. & Lontie, R. (1980). Localisation of the domains in the polypeptide chains of βc-haemocyanin of *Helix pomatia*. In *Structure, Active Site and Function of Invertebrate Oxygen Carriers* (Lamy, J., ed.). Marcel Dekker, New York. In Press

28. Gielens, C., Lanckriet, M. & Lontie, R. (1980). Isolation of the collar fragment of βc-haemocyanin of *Helix pomatia* after proteolysis with enteropeptidase. In *Metalloproteins* (Weser, U., ed. pp. 81–87.) Thieme, Stuttgart.

29. Siggens, K. & Wood, E. J. (1980) Studies on the biosynthesis of gastropod respiratory pigments. In *Structure, Active Site and Function of Invertebrate Oxygen Carriers* (Lamy, J., ed.). Marcel Dekker, New York. In Press.

30. Van Breemen, J. F. L., Wichertjes, T., Muller, M. F. J., van Driel, R. & van Bruggen, E. F. J. (1975). Tubular polymers derived from *Helix pomatia* β-hemocyanin. *Eur. J. Biochem.* **60**, 129–135.

31. Ghiretti-Magaldi, A., Nuzzolo, C. & Ghiretti, F. (1966). Chemical studies on haemocyanins—I. Amino acid composition. *Biochemistry* **5**, 1943–1951.

32. Hall, R. L., Wood, E. J., Kamerling, J. P., Gerwig, G. J. & Vliegenthart, J. F. G. (1977). 3-O-methyl sugars as constituents of glycoproteins. *Biochem. J.* **165**, 173–176.

33. Roxby, R., Miller, K., Blair, D. P. & van Holde, K. E. (1974). Subunits and association equilibria of *Callianassa californiensis* hemocyanin. *Biochemistry* **13**, 1662–1668.

34. Johnson, M. L. & Yphantis, D. A. (1978). Subunit association and heterogeneity of *Limulus polyphemus* hemocyanin. *Biochemistry* **17**, 1448–1455.

35. Antonini, E. & Chiancone, E. (1977). Assembly of multisubunit respiratory proteins. *Annu. Rev. Biophys. Bioeng.* **6**, 239–271.

36. Schepman, A. M. H. (1975). X-ray diffraction and electron microscopy of *Panulirus interuptus* hemocyanin. Thesis, Groningen.

37. Van den Berg, A. A., Gaastra, W. & Kuiper, H. A. (1977). Heterogeneity of *Panulirus interruptus* hemocyanin. In *Structure and Function of Haemocyanin* (Bannister, J. V., ed.) pp. 6–12. Springer-Verlag, Berlin.

38. Sullivan, B., Bonaventura, J. & Bonaventura, C. (1974). Functional differences in the multiple hemocyanins of the horseshoe crab, *Limulus polyphemus*. *Proc. Natl. Acad. Sci. U.S.A.* **71**, 2558–2562.

39. Schutter, W. G., Van Bruggen, E. F. J., Bonaventura, J., Bonaventura, C. & Sullivan, B. (1977). Structure, dissociation and reassembly of *Limulus polyphemus* hemocyanins. In *Structure and Function of Haemocyanin* (Bannister, J. V., ed.), pp. 13–21. Springer-Verlag, Berlin.

40. Lamy, J., Lamy, J., Sizaret, P. Y., Maillet, M. & Weill, J. (1977). Ultrastructure of 16S substances obtained by reassociation using different combinations of three isolated subunits of scorpion hemocyanin. *J. Mol. Biol.* **118**, 869–875.

44	E. J. WOOD

41. Schneider, H.-J., Markl, J., Schartau, W. & Linzen, B. (1977). Subunit heterogeneity of *Eurypelma* haemocyanin and separation of polypeptide chains. *Hoppe-Seyler's Z. Physiol. Chem.* **358**, 1131–1141.
42. Linzen, B. (1980). Structural and functional studies in spider hemocyanin. In *Structure, Active Site, and Function of Invertebrate Oxygen Carriers* (Lamy, J., ed.). Marcel Dekker, New York. In press.
43. Markl, J., Strych, W., Schartau, W., Schneider, H.-J., Schöberl, P. & Linzen, B. (1979). Comparison of the polypeptide chains of *Eurypelma californicus* hemocyanin. *Hoppe-Seyler's Z. Physiol. Chem.* **360**, 639–650.
44. Magnus, K. A. & Love, W. E. (1977). Crystals of a functional 79 000 molecular weight subunit of hemocyanin from *Limulus polyphemus*. *J. Mol. Biol.* **116**, 171–173.
45. Loewe, R., Schmid, R. & Linzen, B. (1977). Subunit association and oxygen-binding properties in spider hemocyanins. In *Structure and Function of Haemocyanins* (Bannister, J. V., ed.) pp. 50–54. Springer-Verlag, Berlin.
46. Van Breemen, J. F. L., Ploegman, J. H. & Van Bruggen, E. F. J. (1979). Structure of *Helix pomatia* oxy-β-hemocyanin and deoxy-β-hemocyanin tubular polymers. *Eur. J. Biochem.* **100**, 61–65.
47. Van Driel, R. & Van Bruggen, E. F. J. (1974). Oxygen-linked association-dissociation of *Helix pomatia* hemocyanin. *Biochemistry* **13**, 4079–4083.
48. Redmond, J. R. (1968). The respiratory function of hemocyanin. In *Physiology and biochemistry of haemocyanin* (Ghiretti, F., ed.), pp. 5–23. Academic Press, London and New York.
49. Brix, O., Lykkeboe, G. & Johansen, K. (1979). Reversed Bohr and Root shifts in hemocyanin of the marine prosobranch, *Buccinum undatum*. *J. Comp. Physiol.* **129**, 97–103.
50. Root, R. W. (1931). The respiratory function of the blood of marine fishes. *Biol. Bull.* **61**, 427–456.
51. Lee, D. L. & Smith, M. H. (1956). Haemoglobins of parasitic animals. *Exp. Parasitol.* **16**, 392–424.
52. Weber, E., Steigemann, W., Jones, T. A. & Huber, R. (1978). The structure of oxy-erythrocruorin at 1·4 Å resolution. *J. Mol. Biol.* **120**, 327–336.
53. Harrington, J. P., Pandolfelli, E. R. & Herskovits, T. T. (1973) *Biochim. Biophys. Acta* **328**, 61–73.
54. Wood, E. J. & Gullick, W. J. (1979). *Planorbis corneus* haemoglobin. Circular dichroism and susceptibility to proteases. *Biochim. Biophys. Acta.* **576**, 456–465.
55. Wood, E. J., Mosby, L. J. & Robinson, M. S. (1976). Characterization of the extra-cellular haemoglobin of *Haemopis sanguisuga*. *Biochem. J.* **153**, 589–596.
56. Chung, M. C. M. & Ellerton, H. D. (1979). The physico-chemical and functional properties of extracellular respiratory haemoglobins and chlorocruorins *Prog. Biophys, Mol. Biol.* **35**, 53–102.
57. Roche, J., Bessis, M. & Thiery, J.-P. (1960). Étude de l'hémoglobin plasmatique de quelques annélids au microscope électronique. *Biochim. Biophys. Acta* **41**, 182–184.
58. Terwilliger, R. C., Terwilliger, N. B., Schabtach, E. & Dangott, L. (1977). Erythrocruorins of *Euzonus micronata* Treadwell: evidence for a dimeric annelid extracellular hemoglobin. *Comp. Biochem. Physiol.* **57A**, 143–149.
59. Guerritore, D., Bonacci, M. L., Brunori, M., Antonini, E., Wyman, J. & Rossi-Fanelli, A. (1965). *J. Mol. Biol.* **13**, 234–237.
60. Di Stefano, L., Mezzasalma, V., Piazzese, S., Russo, G. C. & Salvato, B. (1977). The subunit structure of chlorocruorin. *FEBS Lett.* **79**, 337–339.

61. Vinogradov, S. N., Hall, B. C. & Shlom, J. M. (1976). Subunit homology in invertebrate hemoglobins: a primitive heme binding chain? *Comp. Biochem. Physiol.* **53B**, 89–92.
62. Vinogradov, S. N. Shlom, J. M., Hall, B. C., Kapp, O. S. & Mizukami, H. (1977). The dissociation of *Lumbricus terrestris* hemoglobin: a model of its subunit structure. *Biochim. Biophys. Acta* **492**, 136–155.
63. Garlick, R. L. & Riggs, A. (1979). Partial amino acid sequence of chain AIII of earthworm haemoglobin. *Fed. Proc Fed. Am. Soc. Exp. Biol.* **38**, 343.
64. Harrington, J. P., Suarez, G., Bergese, T. A. & Nagel, R. L. (1978). Subunit interactions of *Glycera dibranchiata* hemoglobin. *J. Biol. Chem.* **253**, 6820–6825.
65. Weber, R. E. & Bol, J. F. (1976). Heterogeneity and oxygen equilibria of haemoglobin from the bloodworm *Glycera gigantea*. *Comp. Biochem. Physiol.* **53B**, 23–30.
66. Allen, D. W. & Wyman, J. (1952). The oxygen equilibrium of hemoglobin of *Arenicola cristata*. *J. Cell. Comp. Physiol.* **39**, 383–389.
67. Weber, R. E. (1975). Respiratory properties of haemoglobins from eunicid polychaetes. *J. Comp. Physiol.* **99**, 297–307.
68. Wells, R. M. G. & Dales, R. P. (1975). In *Proc. IXth Eur. Marine Biol. Symp.* (Barnes, H., ed.) pp. 673–681. Aberdeen University Press.
69. David, M. M. & Daniel, E. (1973). Expression of cooperative oxygen binding at subunit level in earthworm erythrocruorin. *FEBS Lett.* **32**, 293–295.
70. Weichelman, K. J. & Parkhurst, L. J. (1972). Kinetics of ligand binding in the hemoglobin of *Lumbricus terrestris*. *Biochemistry* **11**, 4515–4520.
71. Weichelman, K. J. & Parkhurst, L. J. (1973). Kinetic origin of the pronounced Bohr effect in an annelid hemoglobin. *Biochem. Biophys, Res. Commun.* **52**, 1199–1205.
72. Wood, E. J., Cayley, G. R. & Pearson, J. S. (1977). Oxygen binding by the haemocyanin from *Buccinum undatum*. *J. Mol. Biol.* **109**, 1–11.
73. Wood, E. J. & Mosby, L. J. (1975). Physicochemical properties of *Planorbis corneus* erythrocruorin. *Biochem. J.* **149**, 437–445.
74. Terwilliger, N. C., Terwilliger, R. C. & Schabtach, E. (1976). The quaternary structure of a molluscan (*Heliosoma trivolvis*) extracellular haemoglobin. *Biochim. Biophys. Acta* **453**, 101–110.
75. Figueredo, E. A. Gomez, M. V., Heneine, I. F., Santos, I. O. & Hargreaves, F. B. (1973). Isolation and physicochemical properties of the hemoglobin of *Biomphalaria glabrata* (Mollusca, Planorbidae). *Comp. Biochem. Physiol.* **44B**, 481–491.
76. Read, K. R. H. (1966). Molluscan hemoglobin and myoglobin. In *Physiology of Mollusca* (Wilbur, K. M. & Yonge, C. M. eds.), Vol. 2, pp. 298–232. Academic Press, New York and London.
77. Terwilliger, R. C. & Terwilliger, N. B. (1977). The hemoglobins of the mollusc *Heliosoma trivolvis:* comparison of the radular muscle myoglobin and vascular hemoglobin subunit structures. *Comp. Biochem. Physiol.* **58B**, 283–289.
78. Bonner, A. G. & Laursen, R. A. (1977). The amino acid sequence of a dimeric myoglobin from the gastropod mollusc, *Busycon canaliculatum* L. *FEBS Lett.* **73**, 201–203.
79. Geraci, G., Sada, A. & Cirotto, C. (1977). Cooperative, low-molecular-weight dimeric myoglobins from the radular muscle of the gastropod mollusc *Nassa mutabilis* L. *Eur. J. Biochem.* **77**, 555–560.
80. Terwilliger, N. B. & Terwilliger, R. C. (1978). Oxygen binding domains of a clam (*Cardita borealis*) extracellular hemoglobin. *Biochim. Biophys. Acta* **537**, 77–85.

81. Djangmah, J. S., Gabbott, P. A. & Wood, E. J. (1978). Physico-chemical characteristics and oxygen-binding properties of the multiple haemoglobins of the West African blood clam *Anadara senilis* (L). *Comp. Biochem. Physiol.* **60B**, 245–250.

82. Ilan, E. & Daniel, E. (1979). Structural diversity of arthropod extracellular haemoglobins *Comp. Biochem. Physiol.* **63B**, 303–308.

83. Keilin, D. & Wang, Y. L. (1947). Haemoglobin of *Gastrophilus* larvae. Purification and properties. *Biochem J.* **40**, 855–866.

84. Sladic-Simic, D., Kleinschmidt, T. & Braunitzer, G. (1977). Die Sequenz eines dimeren Hämoglobins (Komponente VII B, *Chironomus thummi thummi*, Diptera). *Hoppe-Seyler's Z. Physiol. Chem.* **358**, 591–594.

85. Gilchrist, B. M. (1954). Haemoglobin in *Artemia*. *Proc. R. Soc. B* **143**, 136–146.

86. Vos, J., Bernaerts, F., Gabriels, I. & Decleir, W. (1979). Aerobic and anaerobic respiration of adult *Artemia salina* L. acclinated to different oxygen concentrations. *Comp. Biochem. Physiol.* **62A**, 545–548.

87. Bowen, S. T., Liebherz, H. G., Poon, M.-C., Chow, V. H. S. & Grigliatti, T. A. (1969). The hemoglobins of *Artemia salina*—I. Determination of phenotype by genotype and environment. *Comp. Biochem. Physiol.* **31**, 733–747.

88. Moens, L. & Kondo, M. (1976). The structure of *Artemia salina* haemoglobins. *Eur. J. Biochem.* **67**, 397–402.

89. Ilan, E. & Daniel, E. (1979). Haemoglobin from the tadpole shrimp, *Lepidurus apus lubbocki*. *Biochem. J.* **183**, 325–330.

90. Ar, A. & Schejter, A. (1970). Isolation and properties of the hemoglobin of the clam shrimp, *Cyzicus* cf. *hierosolymitanus*. *Comp. Biochem. Physiol.* **33**, 481–490.

91. David, M. M., Schejter, A., Daniel, E., Ar, A. & Ben-Shaul, Y. (1977). Subunit structure of hemoglobin from the clam shrimp, *Cyzicus*. *J. Mol. Biol.* **111**, 211–214.

92. Terwilliger, R. C. (1975). Oxygen equilibria and subunit aggregation of a holothurian hemoglobin. *Biochim. Biophys. Acta* **386**, 62–68.

93. Kitto, G. B., Erwin, D., West, R. & Omnaas, J. (1976). N-Terminal substitutions of some sea cucumber haemoglobins, *Comp. Biochem. Physiol.* **55B**, 105–107.

94. Klotz, I. M., Klippenstein, G. L. & Hendrickson, W. A. (1976). Hemerythrin: alternative oxygen carrier. *Science* **192**, 335–344.

95. Loehr, J. S. & Loehr, T. M. (1979). Hemerythrin. A review of structural and spectroscopic properties. *Adv. Inorg. Biochem.* **1**, 235–252.

96. Atkin, C. L., Thelander, L., Reichard, P. & Lang, G. (1973) Iron and free radicals in ribonucleotide reductase. *J. Biol. Chem.* **248**, 7464–7472.

97. Addison, A. W. & Bruce, R. E. (1977). Chemistry of *Phascolosoma lurco* hemerythrin. *Arch. Biochem. Biophys.* **183**, 328–332.

98. Hendrickson, W. A., Klippenstein, G. L. & Ward, K. B. (1975). Tertiary structure of myohemerythrin at low resolution. *Proc. Natl. Acad. Sci. U.S.A.* **72**, 2160–2164.

99. Loehr, J. S., Lammers, P. J., Brimhall, B. & Hermondson, M. A. (1978). Aminoacid sequence of hemerythrin from *Thermiste dyscritum*. *J. Biol. Chem.* **253**, 5726–5731.

100. Ward, K. B., Hendrickson, W. A. & Klippenstein, G. L. (1975). Quaternary and tertiary structure of hemerythrin. *Nature* (*London*) **257**, 818–821.

101. Stenkamp, R. E., Sieker, L. C., Jensen, L. H. & McQueen, J. E. (1978). Structure of methemerythrin at 2·8 Å resolution: computer graphics fit of an averaged electron density map. *Biochemistry*, **17**, 2499–2504.

102. Stenkamp, R. E. & Jensen, L. H. (1979). Hemerythrin and myohemerythrin. A review of models based on X-ray crystallographic data. *Adv. Inorg. Chem.* 1, 219–234.
103. DeWaal, D. J. A. & Wilkins, R. G. (1976). Kinetics of the hemerythrin-oxygen interactions. *J. Biol. Chem.* 251, 2339–2343.
104. Mangum, C. P. & Kondon, M. (1975). The role of coelomic hemerythrin in the sipunculid worm *Phascolopsis gouldi. Comp. Biochem. Physiol.* 50A, 777–785.
105. Schmekel, L. & Weinsche, M. L. (1973). Die Blutdrüse der Doridoidea (Gastropoda, Opisthobranchia) als Ort möglicher Hämocyanin-Synthese. *Z. Morphol. Tiere* 76, 261–284.
106. Messenger, J. B., Muzii, E. O., Nardi, G. & Steinberg, H. (1974) Haemocyanin synthesis and the branchial gland of *Octopus. Nature (London)* 250, 154–155.
107. Decleir, W., Vlaeminck, A., Geladi, P. & Van Grieken, R. (1978). Determination of protein-bound copper and zinc in some organs of the cuttlefish. *Sepia officinalis* L. *Comp. Biochem. Physiol.* 60B, 347–350.
108. Sminia, T. & Boer, H. H. (1973). Haemocyanin production in pore cells of the freshwater snail *Lymnaea stagnalis. Z. Zellforsch. Mikrosk.* 145, 443–445.
109. Sminia, T., Boer, H. H. & Niemantsverdriet, A. (1972). Haemoglobin producing cells in freshwater snails. *Z. Zellforsch. Mikrosk.* 135, 563–568.
110. Fahrenbach, W. H. (1970). The cyanoblast: haemocyanin formation in *Limulus polyphemus. J. Cell. Biol.* 44, 445–453.
111. Moens, L., Vrink, R. & Kondo, M. (1979). *Proc. IX Int. Congr. Biochem. Toronto Abstracts* p. 130.

Liposomes—Bags of Potential

BRENDA E. RYMAN and D. A. TYRRELL†

Department of Biochemistry, Charing Cross Hospital Medical School, Fulham Palace Road, London W6 8RF, England

I. Introduction

The term "liposome" was coined in Alec Bangham's laboratory in Cambridge during the 1960s to describe the concentric multilamellar, bilayered structures which form upon the addition of excess aqueous phase to dry phospholipids.[1] Bangham's observations paved the way for a new technology to study the structure and function of biological membranes: the model was exploited to investigate membrane permeability,[2] molecular motions within membranes using ESR[3,4] and NMR,[5,6] the interaction of anaesthetics[7] and antibiotics[8] with membranes, the immunological properties of membranes,[9,10]

† Current address: The Radiochemical Centre, Amersham, Buckinghamshire.

as well as lipid–protein interactions,[11] which involved studies on a wide range of membrane-bound enzymes and transport systems.

A new decade brought a further flood of potential applications. Clinical medicine had long been searching for a biodegradable carrier for many therapeutic substances, which are toxic when administered alone, or are degraded in the blood, or fail to "home" to the desired target tissue. In 1971 Gregoriadis, Leathwood & Ryman[12] showed that enzymes could be encapsulated in liposomes. They proposed that the complexes might be useful for the *in vivo* delivery of enzymes in the replacement therapy of certain inherited enzyme deficiency diseases such as the glycogen storage diseases. This idea was extended to include the *in vivo* delivery of other therapeutic agents, including drugs (particularly in cancer chemotherapy), chelating agents and hormones. The mechanisms of the *in vitro* interaction of liposomes with cultured cells was intensely studied in the hope of obtaining useful information for the design of the *in vivo* carrier.[13, 14]

Concurrent with the crop of biomedical applications the technology of liposome preparation has made significant advances with the characterization of new classes of vesicles, proving in its turn a headache for investigators with respect to liposome nomenclature![15] Liposomes have become the basis of many interdisciplinary projects involving not only the physical chemists who pioneered the field but also biochemists, cell biologists, pharmacologists, immunologists, physiologists and clinicians.

The field is too large for this essay to be fully comprehensive and the latter part will be biased towards the principal interests of the authors, namely the use of liposomes as therapeutic carriers. However, consideration will be given to the available technology for the production of different classes of liposomes, together with an indication of which types are of most use in any particular field of study. The *in vitro* applications of the technique in membrane biology, immunology and studies of liposome–cultured cell interactions will be dealt with briefly but many excellent fuller reviews will be quoted. The therapeutic aspects will provide the remainder of the text, although much of this field is still largely speculative and many of the ideas may never find clinical application. The major barrier to be overcome is the targetting of liposomes to particular cells and organs of the body (see Section V).

II. Structure and Preparation of Liposomes

A. MULTILAMELLAR VESICLES

When an aqueous phase is added to a dry lipid film of suitable composition, e.g. egg phosphatidylcholine, and gently agitated, the lipid swells and forms completely closed structures consisting of many concentric bilayers of lipid with aqueous phase between the layers. These bodies are termed multilamellar vesicles or MLV, in the generally agreed terminology. They may range in size from 0·1 to 5 μm diameter, and the concentric pattern may be observed by electron microscopy (Fig. 1(a)).

An alternative means of production of MLV, often used in studies of membrane–protein interaction with liposomes, is the detergent dialysis method. Protein and lipid are mixed in the presence of detergent which is subsequently removed by exhaustive dialysis. A number of proteins such ATPase,[16] cytochrome oxidase[17] and human transplantation antigens[18] have been incorporated into liposomes by this general method.

B. SMALL UNILAMELLAR VESICLES

Small unilamellar vesicles (SUV) consist of a single lipid bilayer and have a diameter of 25 nm which, from considerations of geometry and packing of commonly used phospholipids, is the smallest possible vesicle.[19] They may be prepared by exhaustive sonication of MLV using a probe or bath-type sonicator. Any contaminating MLV can be removed by ultracentrifugation or by column chromatography on Sepharose 2B or 4B.[20, 21]

An alternative method for SUV production, avoiding the problems of sonication, was described by Batzri & Korn.[22] It involves injection through a fine hypodermic needle of an ethanolic solution of lipid into an aqueous phase. SUV are formed in very high yield but the use of ethanol has posed a restriction on the application of this technique for several biological applications where labile substances are often involved. Small unilamellar liposomes have also been produced using a French press, when vesicles of diameter 30–50 nm are formed.[23] A negatively stained electron micrograph of SUV is shown in Fig. 1(b).

C. LARGE UNILAMELLAR VESICLES

In 1975 Papahadjopoulos described a class of vesicles having a size range similar to MLV but bounded by a single lipid bilayer.[24] These large unilamellar vesicles (LUV) could only be formed using acid phospholipids such as phosphatidylserine. The preparation requires formation of SUV by sonication and their subsequent fusion into large sheets or cochleate cylinders by the addition of calcium ions. Treatment with EDTA opens out the sheets and LUV are formed.

Deamer described an alternative method, for LUV production,[25, 26] in which ethereal solution of lipid is injected into an aqueous phase maintained at 60°C. The ether evaporates and LUV are formed. The LUV have been characterized by electron microscopy both by negative staining (Fig. 1(c)) and freeze-fracture techniques (Fig. 1(d)).

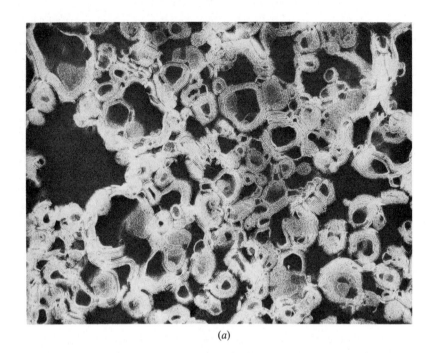

(a)

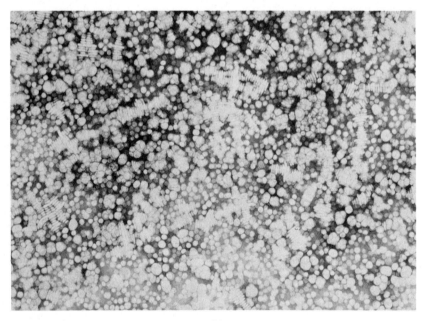

(b)

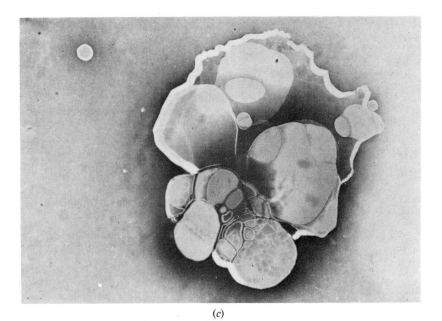

(c)

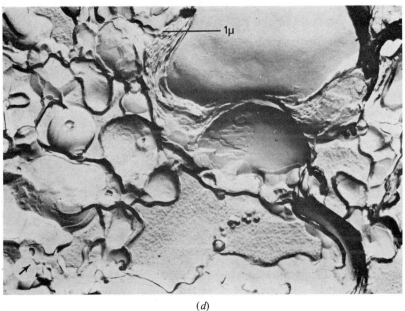

(d)

Fig. 1. Electron microscopy of liposomes. (*a*) Multilamellar vesicles (MLV), negatively stained with phosphotungstic acid (×42 000). (*b*) Small unilamellar vesicles (SUV), negatively stained (×42 000). (*c*) Large unilamellar vesicles (LUV), negatively stained (scale as in (*d*). (*d*) Large unilamellar vesicles (freeze fracture).

LUV have also been prepared by a petroleum ether vaporization method[27] as well as by a detergent dialysis method.[28, 29]

D. REVERSE PHASE VESICLES

The demand for vesicles with a high degree of aqueous phase encapsulation for biological application has recently led to the development of a further class of vesicle.[30] Large unilamellar and oligolamellar vesicles are formed when an aqueous buffer is introduced into a mixture of phospholipid and organic solvent and sonicated, with subsequent removal of the organic solvent by evaporation under reduced pressure. Sonication of the buffered aqueous phase in the organic solvent in the presence of phospholipid molecules probably produces small water droplets stabilized by a phospholipid monolayer. The droplets collapse into a viscous gel-like state when the organic phase is removed. Some of the inverted micelles disintegrate releasing their encapsulated material. However, the excess lipid present helps form a complete bilayer around the remaining micelles, resulting in the formation of vesicles. The nature of the formation of such vesicles has led to their being named reversed phase vesicles (REV).

III. Properties of Liposomes

A. ENTRAPMENT OF MATERIALS IN LIPOSOMES

Substances may be incorporated into liposomes in a number of ways as listed in Table 1. The various mechanisms of incorporation are not mutually exclusive and it is often difficult to prove the precise contributions of each factor. The mode of incorporation of a given material into a liposome may or may not be important. Often the ultimate aim is to achieve maximum incorporation by any available means. However, as an example to the contrary, in studies of the mechanisms of interaction of liposomes with cultured cells it is necessary to incorporate markers which are known to represent the fate of either aqueous or lipid phase alone.

TABLE 1

General methods for incorporating substances into liposomes

1. Aqueous phase entrapment of water-soluble material
2. Electrostatic interaction of charged material with charged lipids
3. Hydrophobic interaction of material with lipids
4. Chemical coupling to surface lipids
5. Any combination of 1–4

Aqueous soluble materials are added to the aqueous phase in liposome preparation; lipid-soluble materials are added to or dried down with the lipid phase in organic solvent. The efficiency of entrapment is dependent upon a number of factors such as type of vesicle, lipid composition of vesicle, charge of vesicle, aqueous buffer strength and the nature of the material to be entrapped. The aqueous trapping efficiencies of the types of vesicle described in Section II is generally REV > LUV > MLV > SUV. Lipid phase entrapment is generally higher and less dependent upon the type of vesicle. An immense variety of materials has been incorporated into liposomes (Section IV and also refs. 31, 32).

Many different types of lipids have been used to prepare liposomes. Those listed in Table 2 are commercially available, and the use of new chemically synthesized lipids for this purpose is being explored.[33, 34] It is important that liposomes are prepared above their transition temperature (i.e. the temperature above which the acyl chains of the lipids are fluid) so that relatively high temperatures are sometimes needed.

TABLE 2

Lipids commonly used in liposome preparation, with their relevant properties

Lipid	Comments
Egg phosphatidylcholine (lecithin)	Most commonly used major component
Dipalmitoylphosphatidylcholine	Synthetic fully saturated phospholipids
Distearoylphosphatidylcholine	Less permeable to aqueous phase than egg lecithin
Sphingomyelin	Often preferred to egg lecithin in immunological studies. Improves liposome stability *in vivo*
Cholesterol	Reduces permeability of egg lecithin vesicles. Maximum incorporation of 50 mole % with phospholipids
Stearylamine	Imparts net +ve charge. Not naturally occurring; may be toxic to cells
Dicetylphosphate	Imparts net −ve charge. Not naturally occurring; may be toxic to cells
Phosphatidic acid	Imparts net −ve charge
Phosphatidylserine	Imparts net −ve charge
Cardiolipin	Antigenic lipid used in immunological applications
Phosphatidylethanolamine	Does not form enclosed vesicles on its own. Useful for coupling materials to the external surface of liposomes. Substituted derivatives used in immunological studies
Lysophosphatidylcholine (lysolecithin)	Increases liposome permeability. May enhance liposome fusion with cells

In any liposome preparation a proportion of the material to be encapsulated will remain unassociated, and this is usually removed by one or more of a variety of techniques such as dialysis, gel filtration or centrifugation.

Liposomally encapsulated material of high molecular weight (>1000) or material associated with the lipid phase does not normally leak out. The preparations are stable for long periods at 4°C and may be sterilized either by γ-irradiation[35] or, in the case of smaller vesicles, by Millipore filtration. Low molecular weight materials may leak out of liposomes, a phenomenon which is often increased in biological environments, such as the presence of serum proteins.[36] The rate of leakage may be reduced by binding of drugs to macromolecules before entrapment.[37] It needs emphasizing that for each new compound entrapped in liposomes, factors such as the nature and degree of entrapment and the amount of leakage under the proposed conditions of use require investigation.

B. LIPOSOMES IN MODEL MEMBRANE STUDIES

In the early 1960s Bangham observed the swelling of egg phosphatidyl-choline in water under the light microscope. When the aqueous phase was altered osmotically, reversible shape changes occurred giving rise to the notion that water was moving across a semi-permeable membrane and that a closed membrane system was under study. This was the birth of the use of liposomes as model membranes. Light microscopic evidence fitted in with that obtained by the electron microscope.[38] Subsequent experiments eliminated the possibility that the changes were electrostatically caused and showed that valinomycin facilitated the diffusion of K^+ over Na^+ from such structures containing equal concentrations of the two ions, proving beyond reasonable doubt the closed nature of the liposome system.[39] Here was a system tailor-made for membrane biologists and the following is an outline of some of the trends of experimentation which have taken place over the past 15 years.

Membrane permeability studies were the obvious first step to determine the leakage of any new substance entrapped. A wide variety of lipid compositions have been investigated for their permeability to ions and electrolytes.[40,41] The effects of steroids,[42] detergents,[43] and local anaesthetics[44] on such permeability have been studied.

The nature of the lipid interactions in liposomes have been examined with the use of NMR,[5,6] ESR[3,4] and fluorescence.[45] These techniques[46] and differential scanning calorimetry[47] are still proving useful in studies of the rates of motion of various membrane components. It is believed that natural membranes contain areas of bilayer which are implicated in fundamental cellular processes such as membrane fusion, so that the fluidity of artificial membranes is an important topic. Recently Chapman has used homogeneous catalysis to modulate the fluidity of liposome membranes.[48]

Proteins play an important role in the stucture of biological membranes, so the effect of various proteins on leakage was studied, and the area of liposome–protein interactions grew into a huge field. Many enzymes have been incorporated into liposomes of varying compositions and their catalytic properties in a lipid environment examined, e.g. β-hydroxybutyrate dehydrogenase,[49] cytochrome oxidase,[17, 50, 51] acetylcholinesterase,[52] urease,[53] acylcoenzyme A : cholesterol acyltransferase.[54] Membrane proteins such as the major erythrocyte glycoprotein, glycophorin, have been incorporated into various types of vesicles and their transmembrane orientation probed in some detail.[55, 56] The current trend is towards the incorporation of complete biological systems into liposomes in a lipid environment similar to that present in vivo. For example, there have been reports of the functional incorporation into liposomes of a mitochondrial H-transporting system,[57] a glucose transport system,[58–61] a microsomal hydroxylase system[62] and Na^+/K^+ ATPase.[63]

Liposomes have also been useful in the study of phospholipid-exchange proteins as membrane probes (for a brief review see refs. 64, 65). It is now thought that such proteins function by redistributing newly synthesized phospholipids from their synthetic loci to subcellular organelles and the plasma membrane.

Liposomes have been used to investigate phenomena which are mediated at the cell surface. Liposomes with various antigenic specificities have been prepared by the incorporation of galgliosides,[66, 69] transplantation antigens[18, 70] or proteins such as glycophorin.[55, 56] Their interactions with lectins can be used to demonstrate the specificity of the liposomes and the results of these experiments may have implications in the targetting liposomes to specific cells (see Section V).

C. LIPOSOMES IN IMMUNOLOGY

Membrane phenomena are extremely important in immunology and liposomes have been useful in investigating a wide range of immunological events. These include the lysis of foreign cells caused by membrane damage, the mediation of such lytic damage by complement and the ability of membranous antigens to induce the formation of humoral antibodies and/or cytotoxic effector cells. In this section some of the more recent work is briefly outlined.

Many naturally occurring lipids, e.g. cardiolipin, have antigenic properties (for a review see ref. 71), but it was some time before the importance of the bilayer structure in producing antibodies to lipids was realized. Kinsky et al. have used several unique N-substituted derivatives of phosphatidylethanolamine such as N-(2,4-dinitrophenyl)phosphatidylethanolamine (DNP-PE) or N-(2,4-dinitrophenyl-ε-aminocaproyl)phosphatidylethanolamine (DNP-cap-PE) which are antigenic.[9, 72] The substances, when incorporated into the lipid bilayer and administered in Freund's complete adjuvant subcutaneously to mice, increased the immune response 16–32-fold with respect to the antibody titre.[73] A synthetic antigen, azobenzenearsonyl-tyrosylphosphatidylethanolamine (ABA-tyr-PE), produces a cell-mediated immune response but no humoral response. However, if it is incorporated into liposomes both types of response are observed.[74]

Kinsky in attempting to enhance the immunogenicity of liposome preparations found that hybrid liposomes, containing two different synthetic antigens in the bilayer (DNP-cap-PE and ABA-tyr-PE) gave an improved antibody response in mice to DNP over that produced by liposomes containing DNP-cap-PE alone.[72] As mentioned above, ABA-tyr-PE is involved in cell-mediated responses, i.e. it is a T (thymus derived) cell determinant. In this situation it is presumed that it has stimulated proliferation of T cells having a helper function in the production of an anti-DNP humoral response. The mitogenic lipid A (a B-cell stimulator) can increase the anti-DNP-cap-PE titre,[75, 76] and immunogenicity is improved if the antigenic determinant is extended away from the surface of the lipid bilayer.[77] Recent experiments carried out by Schuster[78] showed that injection into rabbits of liposomes containing lipid A produced an immune response not only against lipid A but also against the individual lipids used to prepare the liposomes.

The composition of liposomes can also have a profound effect on immunogenicity. For the 16–32-fold enhancement of immune response in liposomes described earlier it was essential that the liposomes were prepared from sphingomyelin and not egg phosphatidylcholine (which gave only a two-fold enhancement).[73] There is a striking difference in the transition temperature of these lipids (egg phosphatidylcholine -8 to $15\,^{\circ}$C; sphingomyelin $42\,^{\circ}$C) and it has been found that liposomes from other lipids with high transition temperatures, e.g. distearoylphosphatidylcholine ($55\,^{\circ}$C) are highly effective immunogens.[79] However, cholesterol, which stabilizes lipid bilayers, had no effect on the immunogenicity of liposomes made from a variety of phospholipids.[72]

These experiments show the great advantages of liposomes as an immunological tool—density, number and type of antigenic determinant can be varied, as well as the chemical environment (i.e. the lipid bilayer). The latter has an effect on antigen expression which might well be extrapolated to natural membranes. Physical techniques such as spin labelling may give more information concerning the orientation of the antigens in the membrane and it may be possible to gain further insights into the way in which membrane antigens on foreign cells are recognized.

The work of Kinsky with synthetic lipid antigens was an outgrowth of his initial work on the complement-mediated lysis of liposomes. The liposome system provides an excellent model which parallels the action of complement on cells and the synthetic lipid antigens described above were designed to overcome the problems of isolating and purifying natural lipid antigens in large quantities. Simple experiments showed that antigen-sensitized liposomes released entrapped glucose when treated with the appropriate antibody and complement. Much of the early work in this area is reviewed elsewhere.[7, 14, 80, 81]

The complement-mediated lysis of liposomes has been used to show that lysis involves phospholipid release[82] but not phospholipid breakdown,[83] to detect antibodies to antigenic lipids,[83, 84] to assay complement[85] and to study the inhibition of complement action by retinal.[73] It has been suggested that liposomes sensitized with an appropriate N-substituted PE derivative might be used as immunogens for the production of antibodies with diagnostic capability (e.g., radioimmunoassay of drugs).[86] Hsia[87] has described an assay procedure using a spin label entrapped in liposomes which have synthetic antigens incorporated into the bilayer. In the presence of the appropriate antibody and complement the spin label is released and detected by ESR. Spin labelling methods have also been used to probe the recognition and triggering of complement in lipid membrane systems[88] and it has been shown that the physical state of the membrane lipids modulates the activation of the first component of complement.[89] The acute phase reactant, C-reactive protein, is able to sensitize appropriate liposomes for damage via the primary complement pathway.[90] In addition liposomes have been used as a model for the activation of the alternative pathway of complement[91] and for K-cell mediated attack.[92] Clearly liposomes are of great importance for studying immunological phenomena.

Other immunological aspects of liposomes have possible therapeutic applications (see Sections IV and V). These include the adjuvant properties of liposomes which are observed towards protein antigens,[93–95] the interaction of viruses with liposomes[96] and the implications of lectin interactions with antigen-containing liposomes in the targetting of vesicles to specific cells.[97]

D. LIPOSOME INTERACTION WITH CELLS IN CULTURE

Liposome interaction with cultured cells has been extensively studied over the past few years, both as a tool for examining important questions in cell biology and as an investigation of liposome potential for the delivery of drugs or genetic material to specific cells *in vivo*. The field of liposome–cell interactions is becoming ever more complex, and there are several recent publications which deal with the subject in depth.[13, 14, 98, 99] The following is a summary of current knowledge.

The types of interaction which might occur between liposomes and cultured cells (Table 3) are frequently studied by incubating vesicles (usually containing radiolabelled lipids and/or entrapped material) with cells in monolayer or suspension for a given length of time under defined conditions. Cell-associated radioactivity is then measured. Occasionally electron spin probes or fluorescent probes have been used in place of radioactive tracers. Recovery of vesicle markers in association with cells does not automatically imply vesicle incorporation into cells. Simple adsorption or exchange of components must be examined by carefully defined control experiments. Vesicle-entrapped materials, which do not normally enter cells in their free state, but which modify cell behaviour in some way when they are internalized, often provide a useful means for demonstrating vesicle uptake. Cyclic AMP has been used for this purpose.[100]

TABLE 3

Possible liposome–cell interactions

---

1. Vesicle adsorption to the cell plasma membrane
2. Endocytosis of the vesicle by the cell
3. Fusion of the vesicle with the cell membrane
4. Exchange of vesicle and cell membrane lipids
5. Exchange of other vesicle and cell membrane components, e.g. proteins
6. Any combination of 1–5

An important factor to consider when interpreting results of liposome-cultured cell experiments is the presence or absence of serum in the incubation medium because serum can have profound effects on the integrity of certain liposomes[36] (see also Section V) and may mediate liposome uptake into cells.[36, 101] There is evidence that all types of vesicle–cell interaction mentioned in Table 3 do occur but that the extent of each is dependent upon the many variables involved. Adsorption of the charged vesicles to cell membranes almost certainly occurs, directly mediated by the opposite charges of vesicles and cells.[36, 102] Pagano et al.[103–105] have shown simple adsorption of vesicles to Chinese hamster fibroblasts using electron microscopy. The biochemical evidence produced suggests the involvement of cell surface proteins in the process.[106] There is also evidence for lipid exchange between vesicles and cells.[104, 107]

Vesicle uptake by mouse 3T3 cells was inhibited by the endocytotic inhibitors, azide and cytochalasin B, suggesting that endocytosis was the predominant method of uptake.[108] Only "solid" vesicles (i.e. lipids below their transition temperature) were endocytosed by 3T3 cells, although, SUV appear to be endocytosed by Acanthamoeba Castelonii.[109] In addition Weissman *et al.*[101, 110] have demonstrated the phagocytosis of fluid MLV when coated with heat-aggregated IgG or IgM by human leucocytes and dogfish phagocytes respectively.

Fusion of liposomes with cells has been demonstrated by a variety of techniques. Microscopic evidence has been presented[102, 111, 112] and lysolecithin and Ca^{2+} ions have been implicated.[113] If vesicle contents are free in the cytoplasm after incubation with cells, fusion is strongly suggested. Weinstein *et al.*[114, 115] have developed an elegant technique for this type of experiment. The fluorescence of the dye 6-carboxyfluorescein is self-quenched at the high concentration present in the liposomes which were incubated with cells. Cytoplasmic fluorescence was observed, suggesting fusion of the vesicles with the plasma membrane and release (and dilution) of the dye in the cell (see also ref. 116).

Cultured cells have already provided several useful models to test the efficacy of liposome-entrapped materials. β-Fructofuranosidase entrapped in MLV caused the disappearance of vacuoles of stored sucrose (induced by prior exposure of the cells to the disaccharide) from mouse peritoneal macrophages.[117] Similarly liposomal amyloglucosidase reduced stored glycogen in cultured cells from a patient with Pompe's disease.[118] Drug resistance of tumour cells has been overcome by incubation of the cells with drug-containing liposomes. Such resistance is often a result of the drug being unable to cross the plasma membrane of the malignant cell. Actinomycin D, and the active metabolite of Ara C (the triphosphate, Ara CTP), have thus been introduced into cultured tumour cells and the metabolite shown to have enhanced activity[119].

With the growing interest in genetic engineering, methods are required to introduce high molecular weight DNA and RNA into cells. Using LUV, mRNA[120] and DNA[121] have been entrapped and introduced into cultured cells. The translation products of mRNA have been identified in the recipient cells.[122, 123] This is an exciting new area.

Finally, cultured cells have provided a model in which to begin to study the targetting of liposomes to specific cell types, an area in which significant progress has to be made if liposomes are to fulfil their early promise. Preliminary experiments have been carried out using antisera to various cultured cell lines incorporated into MLV. MLV containing the antisera specific to the cell type under consideration have shown a modest degree of specificity to those cells.[124] Other workers have studied the influence of incorporated cerebrosides on the interaction of liposomes with cells in culture,[125] the lectin-mediated attachment of liposomes to cells[126] and the incorporation of antibodies into liposomes to demonstrate that they acquire immunological specificity.[127] A recent paper reports on the binding of HLA antigen-containing liposomes to two strains of the bacterium *Neisseria catarrhalis*.[128] By immunological methods Shroit & Pagano[129] have shown that when antibodies bind to antigenic phospholipids in liposomes there is a redistribution of lipid induced in the liposome membrane.

E. THE *IN VIVO* FATE OF LIPOSOMES

The first experiments to investigate the fate of liposomes in experimental animals were performed by Gregoriadis & Ryman.[130] They followed the fate of MLV after intravenous injection into rats using radioactive cholesterol in the lipid phase and labelled amyloglucosidase or albumin in the aqueous phase. The markers accumulated rapidly in the liver and spleen of the animals, with only small amounts in other tissues. Subcellular fractionation showed liposome-associated material in the lysosomal fraction,[131] so that it is likely that liposomes behave in a similar way to other colloidal particles, although there is controversial evidence that not all uptake is by the Kupffer cells of the liver, and there is suggested to be varying degrees of liposome uptake by liver parenchymal cells.[14]

Many reports (see ref. 14) have confirmed the early work on liposome uptake *in vivo* using various markers incorporated into the vesicles and various animal models. Juliano & Stamp[132] were the first to realize that liposome size was of importance in determining *in vivo* fate. They showed that SUV were removed from the circulation at a slower rate than were MLV and further work by Sharma *et al.*[133] confirmed that the rate of clearance was LUV > MLV > SUV. In addition the spleen showed a preference for larger liposome types. The work of Scherphof (see Section V) indicates that certain liposome types are broken down in the blood and thus care must be taken in the interpretation of results even when multiple liposome markers are used in *in vivo* experiments.

There have been many attempts to alter the tissue distribution of liposome uptake by variation in composition and size with only minor degrees of success (e.g. ref. 134). More sophisticated modification may be necessary to achieve such alterations (see Section V).

The intravenous route is not the only one which has been studied. Subcutaneous and intramuscular injection of liposomes may have implications for the slow release of entrapped materials. Segal *et al.*[135] and later workers[136] showed that liposome-entrapped material was released more rapidly from SUV than MLV. The actual fate of subcutaneously administered liposomes is not clear. SUV have been shown to be cleared from the injection site to regional lymph nodes.[137] It is assumed that this represents the passage of intact vesicles into the lymphatics, although it is difficult to obtain direct evidence.

The intraperitoneal route has not received much attention, having little application to clinical medicine. However, some studies have been performed using this route of administration in tumour-bearing animals to investigate the efficacy of liposome-entrapped drugs against ascites tumours.[138] The actual breakdown and redistribution of liposomal lipids and entrapped materials from the peritoneal cavity is not known.

Finally, the oral route of administration of liposomes has generated a good deal of interest with the suggestion that liposomes may be able to enhance the intestinal absorption of substances which are not normally absorbed.[139, 140] This will be discussed in more detail in Section IV.

IV. Therapeutic and Diagnostic Applications of Liposomes

The possible use of liposomes as carriers of materials of therapeutic importance is a subject which has been widely explored over the last ten years, and the many papers have been reviewed.[15, 141-143] We plan to give an outline of the various approaches that have been made, together with discussion of the recent literature which has not been covered by us in earlier reviews.[14, 31]

A. ENZYME REPLACEMENT THERAPY

The earliest suggestion of a therapeutic application for liposomes was for the possible treatment of genetically-inherited storage diseases where the delivery of the missing enzyme to the defective tissue would obviously be of great therapeutic advantage.[12, 144] At the time the method seemed a much more hopeful means of replacement than the then current approach of intravenous infusion of enzyme. The liposome–enzyme complex was shown to be swiftly removed from the circulation after intravenous injection,[130] to be localized in liver and spleen, and to gain access to the lysosomes of liver and to strongly phagocytic cells such as macrophages.[131] In cell culture, several conditions in which stored material has accumulated (e.g., fibroblasts from animals or patients with storage diseases or suitably manipulated normal cells) have been "treated" by liposomally-entrapped enzyme.[117, 118, 145] Cohen et al.[101] claim that coating the liposomes with heat-aggregated immunoglobulins aids the uptake into leukocytes from Tay Sachs patients. The effect of vesicle size on uptake into macrophages has been studied by Schneider (unpublished) who has shown that LUV of size greater than 200 nm diameter are taken up much more readily than SUV by the phagocytic process and that the presence of phosphatidylserine in the vesicle structure facilitates the uptake.

Cystinosis is a further example of a lysosomal storage disease in which cystine accumulates in lysosomes, although the exact defect—possibly a missing enzyme—is not known. Butler et al.[146] have demonstrated that cysteamine-containing liposomes are effective in mobilizing stored cystine in fibroblasts from a patient with this disease and Millard[147] has also considered liposomes in designing strategies to remove cystine from liver lysosomes in cystinosis.

On the question of sequestration of the enzyme from immunological surveillance, the position is complicated. We have already mentioned that liposomes have an adjuvant effect on injected protein.[93, 94, 148] Allison & Gregoriadis[149] showed that entrapment prevents adverse hypersensitivity reactions to the antigen by shielding it from interaction with antibodies. This suggests that administration of entrapped enzyme might circumvent difficulties which may arise from the multiple administrations likely to be necessary for a patient with a genetically inherited disease. However, Hudson et al.[150] demonstrated that protein-loaded liposomes administered intraperitoneally to mice cause a cellular immune response. Therefore, despite the fact that Heath[138] observed a very low adjuvant effect when albumin-containing liposomes were injected *intravenously* the whole question of immunological response to enzyme-loaded liposomes requires further investigation.

There are two reports in which enzyme-containing liposomes have been used in the treatment of patients with lysosomal storage diseases. In 1976, an eight-month-old child with terminal Type II glycogen storage disease (Pompe's disease) was treated with liposomes containing glucoamylase from *Aspergillus niger*.[151] A reduction in liver size and stored liver glycogen occurred, but the massive glycogen accumulation in skeletal and cardiac muscle was not affected. The second case involved an adult patient with Gaucher's disease who was treated over a period of 13 months with liposomes containing glucocerebroside β-glucosidase.[6152] The patient showed reduction in liver size and a relief of pressure symptoms in the abdomen.

It seems unlikely that enzyme delivery employing the liposome carrier will find wide application, because injected liposomes are taken up rapidly by liver and spleen, and at the present time targetting to other tissues such as muscle or brain, which are frequently involved in lysosomal storage diseases, has not been achieved. The studies on the two patients illustrate the possible restricted use of the liposome carrier in disorders where liver and spleen are involved. However, a full "cure" of disorders by this approach awaits advances in liposome technology and targetting. It is possible that gene manipulation may be a more suitable solution to enzyme replacement therapy.

There are further reviews to work discussed in this section.[13, 153–155]

B. CANCER CHEMOTHERAPY

In the last ten years well over 100 different substances have been entrapped in or associated with liposomes, and the drugs used in cancer chemotherapy have contributed very much to the list. Anti-cancer drugs investigated in this way include actinomycin D, 8-azaguanine, bleomycin, BCNU, *cis*-dichloro-biscyclopentylamine platinum, cytosine arabinoside (Ara C), daunomycin, 5-fluorouracil, 6-mercaptopurine, methotrexate, melphalan and vinblastine. References to the many papers on these drugs and other substances can be found in recent articles, some of which are discussed in this section.[14, 31, 156]

The reason why liposome anti-cancer drug research has attracted so much attention is that many drugs used in cancer chemotherapy are highly toxic to normal tissues, particularly bone marrow, kidney, heart and the rapidly dividing cells of the gut lining. Much effort has been directed towards increasing selectivity of anti-cancer drugs,[157] but none of the methods has achieved clinical acceptance. The liposome carrier holds several attractions, including its biodegradability, a reduced rate of excretion of entrapped drug and the possibility of targetting by modification of the lipid bilayer (see Section V).

Much of the early work using liposomally entrapped cytotoxic drugs was carried out *in vitro*, and it has been demonstrated that tumour cells in culture are readily penetrated by liposomes.[119] Drugs can be introduced into suitable cultured cells by entrapment in both the lipid and aqueous phase of liposomes and incorporation of antibodies into the liposomes may aid this uptake.[124] Tumour resistance and the possibility of using liposome-entrapped drug to circumvent it has been studied.[157, 158]

Other experiments of relevance to overcoming drug resistance have been carried out using Ara C. Although Ara C is the form of the drug used in anti-cancer therapy it is believed that it is the triphosphate derivative which is the active form, and that the rate-limiting step in its formation, catalysed by cytidine kinase, may have some part to play in resistance developed during Ara C treatment. A possible bypass mechanism whereby Ara CTP (too polar to enter cells as the free drug) has been entrapped in liposomes to give an enhanced effect on L_{1210} cells in culture compared with free Ara CTP has been described.[159] The same laboratory has produced data to show that cholesterol in the liposome structure enhances the anti-tumour activity of Ara C liposomes.[160]

The persistence of a drug in the plasma of animals after intravenous injection of the entrapped drug has been studied. Encapsulation certainly alters the pharmacokinetics of drugs, as shown by work on actinomycin D, Ara C, daunomycin and vinblastine.[161] In all cases there is a prolongation of half-life—for example <5 min for free daunomycin and >150 min for entrapped drug. Such experiments indicate the possibility that liposomes may provide a slow release mechanism for the drug.

Several reports on the reduction of toxicity of anti-cancer drugs which occurs upon their entrapment in liposomes have been made, but in only one instance[162] is there a full study on the therapeutic index of the drug in the free and entrapped form over a range of doses. Kaye and his collaborators[162-164] have shown that actinomycin D entrapped in liposomes is less toxic than the free drug but entrapment results in an overall loss of activity against a highly actonomycin D-sensitive mouse Ridgway osteosarcoma. It is crucial to determine the effect of entrapment of a drug in liposomes in terms of the therapeutic index (TI). The TI of a drug may be determined as the ratio of the LD_{50} to the minimal effective dose (MED). An increase in LD_{50}, if accompanied by a comparable increase in MED, will lead to little change in TI and no advantage in therapeutic efficiency. The results of Kaye et al.[164] are shown in Fig. 2(a). The loss of activity of the entrapped actinomycin D can clearly be seen. The lack of an increase in TI of the liposome-entrapped drug could be explained in the following way. Actinomycin D is most active therapeutically with a short sharp high level of the drug. Other drugs such as Ara C, which is phase-specific, are more active when high levels are maintained over a period of time. The latter type of drug might have an improved TI in a "long circulating" form such as liposomes. Kosloski et al.[66165] observed enhanced chemotherapeutic efficiency of methotrexate in liposomally encapsulated form against a methotrexate-resistant mouse lymphosarcoma, possibly because of slow release of the drug. No assessment of the effect of the entrapment on TI was made. On the other hand, Kaye[162] found that liposomal entrapment of actinomycin D did not overcome the resistance of a subline of the Ridgway osteosarcoma in vivo.[164] It is likely that the molecular basis for resistance differs in the two cases.

Several other experimental tumours such as Ehrlich ascites, L_{1210} leukaemia and AKR-A lymphoma grown as an ascites tumour are destroyed more readily by entrapped drug. Rustum et al.[138] showed such an effect against L_{1210} leukaemia cells using entrapped Ara C. Phosphatidylserine-containing MLV were more effective carriers for Ara C than were SUV or LUV of comparable composition. Unfortunately the entrapped drug proved to be *more* toxic than the free form.[138] Our own observations on methotrexate have also shown increased toxicity when the drug is entrapped.[164] Using mice bearing the Ridgway osteosarcoma the enhanced activity of the entrapped drug against both normal and tumour tissue can clearly be observed (Fig. 2(b)). These results are in contrast to the ones obtained with actinomycin D (Fig. 2(a)). In neither case is there any change in TI.

Kimelberg et al.[166] followed the distribution of methotrexate following its injection intracerebroventricularly in liposomes. The resultant reduced leakage of the drug from the central nervous system to plasma compared with free drug provides a suitable formulation for such a route of injection. The stearylamine in the liposomes in these experiments[166] did not produce any toxic effects,

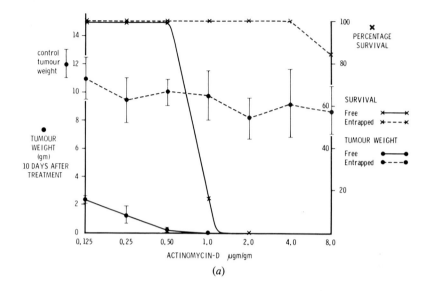

(a)

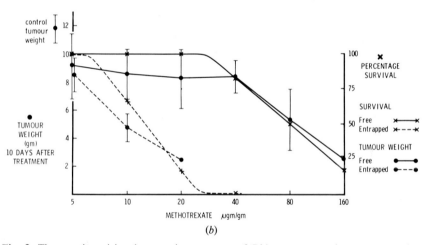

(b)

Fig. 2. Therapeutic-toxicity data on the treatment of Ridgway osteogenic sarcoma-bearing AKR mice with actinomycin D (a) and methotrexate (b) in the free and (cationic) liposome-entrapped form. Liposome composition was phosphatidylcholine:cholesterol:stearyl-amine in the molar ratio 18:4:5. Each point refers to mean for six mice with standard error shown. Untreated control tumour weight at 10 days (11·8 ± 1·0 g) is also shown. For actinomycin D in the free form, the LD_{50} is 0·8 μg/g and the MED_{10} (dose inhibiting tumour growth to 10% of control) is 0·3 μg/g. Neither LD_{50} nor MED_{10} is reached using liposome-entrapped actinomycin D in doses up to 8 μg/g. In contrast, for methotrexate in the free form the LD_{50} is 80 μg/g and the MED_{20} (20% control tumour growth inhibition) is 160 μg/g, whereas for methotrexate in the liposome-entrapped form, the LD_{50} is 12 μg/g and the MED_{20} is 20 μg/g, i.e. there is no overall improvement in the therapeutic index.

although toxicity was observed with 19–38-fold higher doses of stearylamine-containing liposomes.[167]

The use of liposomes to reach lung tissue and hence to possibly deliver drugs to lung tumours or metastases has been considered by Hunt et al.[168] who examined the retention of Ara C in liposomes of differing size in the lungs of mice following i.v. injection. McCullough & Juliano demonstrated the targetting of Ara C to lung tissue by administering it in liposomes as an intratracheal bolus. The entrapped drug is retained better in the tissue, and causes less suppression of bone-marrow and of gut metabolism than the free substance.[169]

A question that arises because of the greater efficacy of liposome-entrapped drugs against tumours is whether the liposomes actually gain access to tumour cells (or macrophages associated with the tumour). Some tumours are phagocytic but there is no clear-cut evidence of liposome uptake. In some, but not all, animal tumours studied, evidence for the uptake of small negatively charged fluid liposomes has been obtained[170, 171], although there is the possibility that the radiolabel is no longer associated with the liposome.

Animal work has been extended to human patients. One of the early investigations in which ^{125}I-labelled albumin was entrapped in the liposome and the radioactivity located at post-mortem showed uptake into tumour tissue.[172] It seems likely that the albumin alone was being taken up because of the lack of localization found in a wide variety of tumours using ^{99m}Tc[173, 174] or ^{111}In-labelled bleomycin[165] loaded liposomes.

Although the drug-loaded liposome is a fascinating possibility in cancer chemotherapy, positive *in vivo* data demonstrating enhanced cytotoxicity is limited to those phase-specific drugs such as Ara C and methotrexate whose activity is clearly dependent upon the duration of drug exposure. Meanwhile, the search for alternative ways of attacking tumour cells continues. Activation of lymphocyte attack employing liposomes has been considered[175] with RNA incorporated into liposomes. These vesicles contained both antibody to lymphocytes and also RNA from guinea pigs immunized with a syngeneic line of hepatocarcinoma cells.

(1) Lymph node imaging

Segal et al.[135] observed that liposomally entrapped drugs can be detected in regional lymph nodes after intratesticular injection. The feasibility of utilizing liposomes to detect the spread of malignant solid tumours to the regional lymph nodes through the lymphatics has been considered.[137, 176] The ability to detect such spread would be of great importance in treatment since at the present time clinical methods and lymphography are inaccurate. Accurate detection of those lymph nodes involved in metastatic spread would allow rationalization of surgical ablation as well as of other treatments.

[99m]Tc-labelled liposomes form a new and promising lymph node imaging agent, small neutral or cationic ones being particularly effective. The ^{99m}Tc localized in the lymph node can readily be visualized by external γ-camera imaging[176] following interstitial liposome administration to normal rats. In an animal model in which a footpad tumour metastasises regularly and reproducibly through the adjacent regional lymph nodes, it has been shown that there is a differential uptake between these lymph nodes and normal ones, and suppression of uptake of the radionuclide is observed in the nodes containing tumour deposits.[177] Encouraging results were obtained when these methods were applied to axillary lymphoscintigraphy in breast cancer.[178] Hopefully accurate assessment of metastases before surgery and without recourse to histological examination may be made. Other conditions in which lymph node involvement in malignant diseases can occur are under investigation.

C. CHELATION THERAPY

Chelating agents such as EDTA, DTPA (diethylenetriaminepentaacetic acid) and desferrioxamine are used intravenously and intraperitoneally for treating metal poisoning, although they do not readily cross cell membranes. Moreover they are rapidly excreted from the body and desferrioxamine is rapidly degraded in the plasma. To overcome some of these difficulties, Rahman and her group[179] investigated the use of liposome-entrapped chelating agents, and showed that plutonium overload can be treated with vesicles containing DTPA. Not only is the accumulation of colloidal plutonium in liver tissue effectively removed by the entrapped DTPA, but the translocation of plutonium from liver to boney skeleton, which occurs in the presence of free DTPA, is prevented. Rahman's group has also investigated the use of liposomes in treatment of lead poisoning, mercury and gold poisoning, and in accumulation of cadmium, copper, calcium and thorium.[179]

Iron overload occurs in both haemosiderosis and haemachromatosis, and treatment includes the use of chelating agents such as desferrioxamine. The possibility of using vesicle-entrapped desferrioxamine rather than free drug has been investigated.[179] Iron overload in animals was produced by injection of either ferritin, which loads hepatocytes, or ^{59}Fe-labelled, slightly damaged, erythrocytes which are rapidly taken up by Kupffer cells of the liver and by the spleen. By observing excretion of ^{59}Fe in the urine of the animals, the efficacy of entrapped compared with free chelating agent was assessed. In the case of the ferritin model, incorporation of glycolipid, which has a high affinity for hepatocytes,[180] into liposome structure resulted in 1–2% more effective removal of iron than when free desferrioxamine was given. SUV were found to be more effective than MLV. In the erythrocyte model entrapment of the chelating agent resulted in a 10% increase in excretion over the control (non-entrapped) treatment; MLV were more effective than SUV. In confirmation Young et al.[183] have shown that an increased urinary excretion of iron occurs in iron-overloaded mice treated with desferrioxamine entrapped in liposomes compared with free desferrioxamine.

In other experiments[181, 182] iron overload was induced by long-term injection of an iron preparation (Jectofer) when accumulation occurred in both hepatocytes and Kupffer cells. The administration of MLV containing either DTPA or desferrioxamine did not result in significant lowering of iron levels in either the livers of these animals or in crude lysosomal fractions isolated from this tissue, lysosomes being the site at which iron accumulation is thought to occur in iron overload conditions. The apparent discrepancy between the various models calls for further investigation. The use of the liposome vesicle in treating metal poisoning is clearly a hopeful approach to improved chelation therapy.

D. CARRIERS OF HORMONES

(1) Cortisol esters

A group from the Strangeways Laboratory, Cambridge, U.K.[184-187] has shown that experimentally induced arthritis in rabbits can be "treated" more effectively when cortisol palmitate is incorporated into liposomes made of dipalmitoylphosphatidylcholine and phosphatidic acid than with the free cortisol ester. A liposome preparation containing 35 μg of the cortisol ester, injected directly into the arthritic joint, led to a clear-cut beneficial effect compared with non-entrapped drug four days after the injection, with increased response up to 100–200 μg of ester.

In a pilot study on six patients with arthritis, 2 mg cortisol ester entrapped in vesicles appeared as effective as 20–25 mg of free drug, which is a normal dose. While this technique of liposome drug delivery into the arthritic joint looks promising, it is perhaps questionable from a therapeutic point of view, because if the patient becomes too free of pain, he/she may aggravate the condition by use of the joint and do further damage. Nevertheless the liposome-entrapped anti-inflammatory agents offers an attractive approach. Entrapment of α_1 anti-trypsin and soya bean trypsin inhibitor in MLV with a view to using the liposomes in vivo to introduce defined protease inhibitors into cells responsible for enzyme release in connective tissue diseases, including arthritis, has been reported by Finkelstein et al.[188]

(2) Insulin

The possibility of using insulin entrapped in liposomes as a means of achieving the long wished for oral preparation of this peptide was indicated in a patent in 1974. The first scientific evidence was published by Patel & Ryman in 1976,[139] who observed in streptozotocin diabetic rats a lowering of blood glucose when the animals were fed insulin entrapped in MLV. The work was confirmed and extended.[31, 137, 189, 190] However, Patel et al.[191] concluded from their studies that the low effectiveness of absorption of insulin entrapped in MLV (about 1% of the dose appears to be absorbed after oral administration) makes it unlikely that liposomally entrapped insulin will be of therapeutic importance, unless this problem is overcome.

This aspect is being studied. Thus Bridges et al.[192] demonstrated that uptake of polyvinylpyrrolidone in everted intestinal sacs prepared from adult rats was increased by a factor of 4 when the polymer was entrapped in MLV composed of dipalmitoylphosphatidylcholine and dipalmitoylphosphatidic acid, although Whitmore & Wheeler[193] failed to observe liposome uptake in a similar system. In an attempt to gain an understanding of the factors which may be involved in transport of drugs across the intestinal wall, Hori et al.[194] incorporated intestinal lipids into vesicle membranes and concluded that the free fatty acids present play an important role in the absorption process.

The feasibility of employing lipid derivatives in the liposome structure which are not vulnerable to attack by gut enzymes, or which yield liposomes not readily disrupted by bile salts, has attracted attention. Richards & Gardner[195] have shown that more fluid liposomes, that is those prepared from lipids with transition temperatures below body temperature, are totally disrupted by bile salts, whereas less fluid ones are less vulnerable to attack.

There is no clear evidence that intact liposomes are absorbed through the gut wall,[185] although the lipid overcoat around the insulin probably protects the polypeptide against proteolytic attack. Liposomes may not yield a new formulation for oral insulin administration because, apart from the low amount absorbed, a dose dependency cannot be demonstrated. Nevertheless, the liposome-insulin orientated work has stimulated new efforts to find non-parenteral preparation for this hormone, e.g. nasal route absorption.[196]

Much work is proceeding in the pharmaceutical industry on new methods of delivery of non-intestinally active drugs other than insulin. For example, two patents[197, 198] appeared in 1978 in which it is claimed that orally active drugs and vaccines can be prepared from materials normally not given by mouth, by suspending them in egg phosphatidylcholine and cholesterol and filling the suspension into capsules coated with hydroxypropylmethylcellulose phthalate. The patent claims included materials such as proteins and peptides. A further patent[199] claims that heparin, another non-intestinally active drug, administered in liposomes and given orally is more effective in increasing bleeding time in mice than heparin in the free state. The outcome of these various patent claims is eagerly awaited in the hope that the results may help in the formulation of a non-injectable preparation of insulin.

Stevenson, Patel and their collaborators[200] have recently been examining the absorption of liposome-bound insulin following its *subcutaneous* administration to diabetic animals, and concluded that the entrapment of the hormone in liposomes results in a preparation which increases the duration of its biological response. Insulin-liposomes administered *intraperitoneally* lead to a fall in blood glucose in diabetic rats but appear to have no advantage over free insulin.[139]

E. TREATMENT OF TROPICAL DISEASES

There are probably more than 1 million deaths per year from the two tropical diseases, malaria and schistosomiasis. There are almost 11–12 million cases of leprosy (caused by a bacterial parasite), and 30% of these patients are left with irreversible changes. Trypanosomiasis (African sleeping sickness) and Chagas disease of South America are likewise tropical diseases which carry a high morbidity rate for man and cattle.

Leishmaniasis prevalence is unknown, but it is probably greater than that of malaria. The visceral form of leishmaniasis, where the causative agent is *L. donovani*, is characterized by the invasion of the reticuloendothelial system, especially the liver and spleen, by the parasite. This led Black et al.[201] to investigate the efficacy of liposomally entrapped drug for treatment of the disease. The drug was Pentostam (Na stibogluconate antimony) and the strategy was based on the known uptake of liposomes into the reticuloendothelial system of the liver and spleen[130]; the entrapped drug was 200 times more effective than the free drug in mice infected with the parasite.

Later New et al.[202] showed that liposomally entrapped antimony drugs are effective against both the visceral and the cutaneous (*L. tropica*) form of leishmaniasis. New (unpublished results) has also shown that primaquine therapy in visceral leishmaniasis results in synergy between liposomes and the drug, an effect possibly related to the enhanced entry of the drug into Kupffer cells or increased permeability of drug into the parasite. This synergism is observed when there is no physical interaction between liposomes and drug and is similar to the observed adjuvant effect of liposomes seen by Heath[148] where again no physical contact between antigen (bovine serum albumin) and liposomes was required to elicit an adjuvant effect.

Alving et al.[203] confirmed the work of Black & New and their colleagues and demonstrated the 700–800-fold greater efficacy of liposomally entrapped antimony drugs in the visceral form of this tropical disease. This group reported fascinating work on the possible use of liposomes in treatment of malaria.[204, 205] Hepatic parenchymal cells are involved in one stage of the life cycle of the malarial sporozoites, and these cells are an attractive target for liposomes. But the efficacy of liposome intervention in malaria lies at a more sophisticated level. Primaquine in liposomes does not increase its therapeutic efficiency, although toxicity is lessened. However, a specific lipid which appears to block parenchymal cell receptors for the parasite has been found.[204, 205] Based on the work of Morell & Ashwell[206] who showed that receptors on parenchymal cells recognize galactose, Alving *et al.*[204, 205] prepared MLV liposomes containing the glycolipids galacto-, glucosyl- or lactosyl-ceramide as the blocking entities. The TI of the preparation is virtually infinite.

A parasitic worm, *Schistosoma mansoni,* invades the mesenteric and hepatic veins, and is also occasionally found in liver sinusoids. Black (unpublished results) was unable to show any improvement in TI as a result of using entrapped antimonial drugs. Trypanosomiasis therapy employing suitable entrapped drug has been reported.[207, 208]

These are exciting developments in the field of tropical medicine and they will undoubtedly be extended.

F. OTHER USES OF LIPOSOMES

(1) Respiratory distress

Respiratory distress of the new-born in which lack of lung surfactant occurs and where, with the "first breath", the surface area of the lung tissue increases dramatically has been considered.[209] This work has indicated that *dry* dipalmitoylphosphatidylcholine, one of the main lung phospholipids, may be helpful in treating respiratory distress. With new methods of assessing the ratio of phosphatidylcholine to sphingomyelin before birth, and hence the means of an early warning system for foetal lung immaturity,[210] the treatment with dry phospholipid may become a reality.

(2) Digoxin overload

Unpredictable life-threatening toxicity still occurs in some patients treated with digoxin and the possibility of using antibodies to counteract the effects of the drug is an attractive approach. There is, however, the problem of long circulation of the resultant immune complexes to be considered. In an effort to overcome this disadvantage, digoxin antibodies were incorporated into MLV and used in animal experiments to demonstrate their effectiveness in neutralizing circulating digoxin. The rapid removal of such MLV by the liver and spleen, because of the natural targetting of vesicles, overcomes the disadvantage experienced when antibody alone is used.[211, 212]

(3) Myocardial infarction

Caride and his colleagues showed in 1977[213-215] that MLV accumulate after intravenous injection at the site of an artificially induced myocardial infarction in animals, leading to the possibility of selective drug or support therapy to the injured tissue. They showed that positively and neutrally charged MLV accumulated most readily, to a large extent in the endocardium. A few liposomes labelled with entrapped horseradish peroxidase could be visualized in capillaries near the myocardial infarction after injection of the MLV, although there was no indication of intact liposomes inside the myocytes. The suggested reasons for the apparent preferential accumulation of the MLV at the site of infarction include increased capillary permeability and myocardial cell damage, changes in blood flow and the destruction of MLV at the ischaemic region of the heart which is enhanced by local tissue changes and necrosis. The impaired lymphatic drainage from the interstitial space of the infarcted myocardium may also contribute to the observed accumulation.

An important observation in connection with possible treatment of myocardial infarction is that negatively charged vesicles promote blood clotting, positive vesicles extend the clotting time, and neutral vesicles have little effect. While positively charged vesicles thus appear most promising, the toxic effects of liposomes containing stearylamine[14, 31] should be remembered.

(4) Adjuvants and virus research

Earlier work by Allison & Gregoriadis[94] and Heath et al.[93, 148] clearly showed that liposomes can behave as adjuvants and that the lipid composition of the vesicles used,[93] the route of administration[148] and the species of animal used[150] influence the response. Lysolecithin as a component of the vesicle impairs the adjuvant effect.[216, 217] The use of liposomes as adjuvants has several advantages, including their completely biodegradable nature and the fact that their lipid components are not foreign.

Virus research has also activated liposome enthusiasts, and this area was recently reviewed by Haywood.[96] Viruses may help us to understand factors which confer tissue specificity/targetting on the liposome, a need particularly felt in the design of liposomes for the various proposed uses which are outlined in this section. Viruses have natural tissue specificity although the mechanism is not understood. For example, the mumps virus targets to the parotid glands and the central nervous system, whereas the Sendai virus invades the respiratory system of human babies and mice. Haywood,[218] in investigating the way in which viruses enter cell membranes, has shown the importance of gangliosides and phosphatidylethanolamine on the binding, fusion and phagocytosis of Sendai virus into liposomes.

Almeida *et al.*[219] showed in 1975 that incorporation of the surface subunits (neuraminidase and haemagglutinin) of the influenza virus on to the lipid bilayers of liposomes yielded a preparation which they suggested could be used as a new form of vaccine. The advantages of such a preparation appear to lie in reduced pyrogenicity compared with whole influenza virus and with maintenance of immunogenicity. The authors[219] coined the name "virosomes". Morein[220] has considered fully such virosome preparations as vaccines against some enveloped animal viruses. Later, Manesis *et al.*[221] prepared a hepatitis B virosome which enhances both humoral and cell-mediated immunity to the antigen. Kramp *et al.*[222] observed liposome enhancement of immunogenicity of adenovirus type 5 hexon and fiber vaccines, and Finberg and his colleagues[223] observed induction of virus-specific cytotoxic T-lymphocytes using solubilized viral and membrane protein in liposomes. There are other studies on the assembly of viral proteins with the phospholipid membranes of liposomes.[224–226]

The virus-liposome field is clearly a very active one and the production of new vaccines seems highly probable.

(5) *The introduction of informational molecules (RNA and DNA) into cells via liposomes*

In 1978 Dimitriadis[227] described the introduction of RNA into vesicles containing phosphatidylserine, which was used to minimize binding of the informational molecule with the phospholipid. He demonstrated that following the introduction of mRNA for globin into lymphocytes the message was read and globin produced.[123] The RNA was thus probably protected by the liposomes from ribonuclease attack. Similarly, globin mRNA in liposomes could be inserted and translated in a human cell line.[122] Wilson *et al.*[228] have recently demonstrated that polio virus RNA entrapped in liposomes, can be introduced into cells not normally infectable by the virus (it is only primate cells which carry the virus receptor). Only 1% of the polio virus RNA became associated with the host cells, but so effective was delivery of the virus that all cells were killed.

Hoffmann *et al.*[121] demonstrated that liposomes containing high molecular weight DNA entrapped in egg phosphatidylcholine vesicles might be useful in the introduction of foreign DNA into cells, and also might serve as a model for membrane interaction. Dimitriadis[229] has subsequently entrapped plasmid DNA in liposomes, and Mukherjee *et al.*[230] have reported that metaphase chromosomes could also be prepared in vesicle-bound form and could, like the plasmid DNA, have potential in gene transfer and enzyme replacement therapy (see Section IVA). Three further papers on DNA transfer have appeared recently.[231–233]

Fraley *et al.*[234] reported that the plasmid pBR322 can be entrapped in liposomes and transferred to *E. coli*: the addition of deoxyribonuclease in the transformation mixture was without effect when liposomes were employed, whereas transformation by free plasmid was totally eliminated after treatment with the enzyme. Mannino *et al.*[231] encapsulated fragments of phage T7 DNA (molecular weight $0.27–14.19 \times 10^6$) into LUV made of phosphatidylserine and again have shown their resistance to degradation by the DNA nucleases.

Although the incorporation of DNA molecules into vesicles has great potential and is of considerable interest, their transfer into the cell nucleus, which is the aim of much research, may pose considerable problems.

(6) Miscellaneous possible applications of liposomes

Liposomes have been used in further studies on anaesthetic action,[233, 235] thermal transitions of phospholipids,[236] neurobiological applications,[237–239] mast cell degradation,[240] calcium regulation,[241] leucocyte phagocytosis,[242, 243] the pathogenesis of Graves ophthalmology,[244] botanical research,[245, 246] drug interactions with membranes,[247] bacterial cell wall biosynthesis,[248] lectin receptor interactions,[249, 250] and microvilli formation in cells.[251]

V. Future Outlook/Targetting

There are clearly many aspects in the liposome field of research which require further investigation before the assessment of the potential of these little fatty vesicles can be made. The two main areas seem to us to be:

(1) distribution and integrity of liposomes in body fluid, and
(2) targetting of liposomes to cells and tissues.

(1) Integrity of liposomes in body fluids

The original observation[130] on the fate of i.v. injected liposomes *in vivo* was made in 1972 using liposomes carrying radioiodinated albumin and with [^{3}H]cholesterol in the lamellar structure of the carrier. The two labels disappeared concomitantly from circulation and the tritium label was apparently evenly spread (as judged by autoradiographic techniques) between the various types of cell in the liver. The conclusion was reached[130] that liposomes remained intact in the circulation over the time period studied and that this rapid removal was associated with uptake by the liver and also by the spleen. However, the presence of serum components influences the uptake of liposomes and their contents into cells,[36, 252] and moreover not all forms of liposomes are stable when in contact with blood components, even though they are stable in buffer solutions for many weeks. Furthermore, we know that cholesterol exchange takes place very rapidly from vesicles to cells and tissue[14] and indeed even occurs with viruses,[253] and therefore the earlier interpretation of these original results with respect to *in vivo* distribution has to be modified.

It is known that the fenestration in the endothelial cells of the liver, through which liposomes need to pass from the circulation if they are to gain access to either Kupffer cells (part of the reticuloendothelial system) or parenchymal liver cells are of the order of 1000 Å (100 nm) and so it may be reasonable to expect that the larger varieties of liposomes described in the earlier sections of this essay will not gain access to liver cells. Furthermore, several other tissues when considered with respect to pore size are unlikely to be candidates for liposome entry unless modifications to the route of access are made. Scherphof and his colleagues have made careful studies of both the related problems of access to liver tissue and the cell types involved and also to the integrity of liposomes. They have shown[254-259] that in the case of SUV containing low levels of cholesterol (20%) the half-life of the vesicles after i.v. injection into animals is approximately 73 min and some 84% of the phosphatidylcholine (PC) in the vesicle is transferred to high-density lipoproteins (HDL). This transfer leads to a rapid solubilization of the liposome with release of its contents. At the same time the process leads to the formation of a lipoprotein particle which donates the lipid to the parenchymal liver cell, provided that the size of the particle is <100 nm. MLV (handshaken), on the other hand, have a half-life of 65 min after injection and with a 20% cholesterol content in their lipid composition there is much less solubilization (about 10%) and transfer of PC to HDL compared with SUV and only a partial release of liposome content occurs. The bulk of these liposomes is taken up by the Kupffer cells. This process of transfer of PC to HDL with release of liposome contents, which has been studied by other groups,[260-262] is clearly an undesirable feature for many of the proposed uses of liposomes and can be considerably reduced if the cholesterol content of the vesicle is increased. For example, Scherphof's group has shown that when a $3:2$ molar ratio of PC:cholesterol or a $1:1$ sphingomyelin:cholesterol is used in the preparation of MLV there is no transfer and only a very small leakage. MLV made with one fatty acid type, e.g. dimyristoylphosphatidylcholine, have also been shown by Scherphof's group to be unstable in the blood after injection and it was also shown that the incorporation of 33% cholesterol in the vesicle structure confers stability and suppresses the loss of contents.[259]

It would therefore seem that MLV with a high cholesterol content may be the most suitable form to attain stability in biological fluids. SUV are much less stable and may not only be solubilized by interaction with HDL but some (e.g., when made of phosphatidylserine) may be lysed by Ca^{2+}.[263] It will, however, be appreciated that the possible combination of lipid composition and size of vesicle is enormous and only by extensive experimentation will it be known which vesicle, if any, is suitable for therapeutic application. The determination of the exact distribution of liposomes into cells and tissues is fraught with difficulties and the selection of suitable "markers" for following liposomes is of paramount importance. The use of at least two markers is recommended. Markers such as radiolabelled lipids, radiolabelled inulin (which entraps well in the aqueous phase), horseradish peroxidase and ferritin can be used, while 6-carboxyfluorescein[114] is an excellent marker for measuring transfer of liposome contents into cells, and the newly described covalent attachment of horseradish peroxidase to liposomes[264] should be very useful in studying vesicle–cell interactions. The possibility that γ-ray perturbed angular correlation studies may assist in determining conditions and localization of vesicles in specific tissues has been suggested by Mauk & Gamble.[265]

The problems of possible toxicity of liposomes will also need to be tackled if the potential of liposomes as a therapeutic carrier system is to become a reality; a brief summary on toxicity has been recently published.[156]

(2) Targetting of liposomes to cells and tissue

In our discussions on the possible therapeutic potential of the liposome carrier in the delivery and targetting of drugs, hormones, enzymes and other materials of therapeutic interest, it must be obvious that one of the several drawbacks is that the carrier is so readily removed by the reticuloendothelial cells of liver and spleen. Although this natural homing of liposomes to liver and spleen may be an advantage in specific cases, e.g. some tropical diseases therapy (see Section IV.E) or removal of digoxin by liposome-digoxin antibody,[211, 212] it has a considerable disadvantage when other cells and tissues are the target in the delivery system. Several approaches have to be made to design strategies to overcome this disadvantage and introduce other "homing" mechanisms.

The manipulation of the liposome structure as far as composition of lipids, degree of fluidity, size of vesicle and net charge on the liposome surface are concerned, has not dramatically changed the predominant role of liver and spleen in liposome uptake after injection into animals or patients. Jonah et al.[180] have shown that some modification in tissue distribution can be achieved by changes in lipid composition and that the introduction of sialogangliosides into the liposome decreased the uptake by liver: Similar effects were obtained by using cardiolipin-containing liposomes.[266] The inclusion of glycoceramides in liposomes, which results in enhanced uptake by the liver, has been referred to earlier. The feasibility of effecting a partial blockade of the reticulo-endothelial system before administration of liposomes has been considered. Gregoriadis & Neerunjun[267] attempted to alter the distribution of intravenously injected liposomes by such means, but no effect was seen. However, in these experiments there was no definition or proof of "blockade" and the dose and timing of the administration of the blockading injection was not established as being appropriate. Souhami & Patel (unpublished results) have used dextran sulphate (mol. wt. 500 000) to produce "blockade" in mice in a dose and at a time in which blockade can readily be demonstrated by the inhibition of uptake of ^{51}Cr-labelled sheep red blood cells. Under these conditions the hepatic uptake of intravenously injected neutral liposomes is reduced and prolonged circulation of the liposomes is seen as a consequence of the depressed hepatic clearance. Spleen uptake is enhanced by "blockade", a result similar to that found with ^{51}Cr-labelled sheep RBC, but occurring to a lesser degree.

Attempts to manipulate the properties of the liposome surface to effect targetting to tumour tissue has led to the general conclusion that small negatively charged liposomes with high fluidity are favoured for apparent localization in animal tumour tissues.[137, 170, 268] While such manipulation appears hopeful in animal tumour targeting, there is regretfully no evidence of tumour uptake in man.[173] The reported "uptake" of liposomes into lung tissue may represent a size factor in that large vesicles may be expected to lodge more easily in small capillary beds.

The attachment of proteins with "recognition" properties is an area which has occupied, and continues to occupy, the thoughts and energies of many investigators. It is perhaps not surprising that immunoglobins of various types are candidates in this respect. Gregoriadis & Neerunjun[124] showed that by incorporating into liposomes antibodies to specific cells (HeLa or fibroblasts), such vesicles would more readily associate with the appropriate cells in culture compared with liposomes containing no antibody. Weissmann and his colleagues have demonstrated that heat-aggregated antibodies on the surface of liposomes resulted in enhanced uptake of the vesicles into dogfish phagocytes.[110] It is thought that the heat aggregation induces a conformational change in the Fc region of the IgG which permits interaction with the liposomes.[269] Such earlier success in the *in vitro* conditions has, however, not been observed *in vivo* and many workers have investigated the binding of immunoglobins to liposomes by various techniques, e.g. sonication.[270] Heath *et al.*[264] have recently reviewed the field of covalent attachment of proteins to liposomes which will no doubt be of assistance to those seeking to bind recognition proteins to lipid membranes. Torchilin *et al.*[271] have reported on the use of "spacers" in protein binding to liposomes. While the attachment of recognition molecules and proteins to the surface of vesicles to induce specific targetting offers considerable promise, there is so far insufficient evidence to suggest that this strategy will be successful.

It will be appreciated that targeting liposomes to key cells and tissues for a variety of therapeutic purposes is a complicated concept. Consider the situation of the anti-cancer drug-loaded liposome with its target the tumour cells. We may envisage the *liposome* as the *letter* bomb which needs to be *posted* with the right *address* (tissue or cell), and needs also to gain *entry* (through the letter box) in order to *deliver* its contents, i.e. to destroy the target. The process therefore has at least three critical barriers to be overcome, and only if all can be surmounted will the delivery system be complete and successful. Considering the complexity of the requirements for success in the intact animal, it is perhaps surprising that any progress has been made, particularly with respect to targetting. Weinstein et al.[272] have illustrated the difficulties in the posting → delivery pathway outlined above in their work on antibody-mediated targetting of liposomes. They have shown[272] that, whereas liposomes will bind to lymphocytes, such binding does not lead to the entry of the vesicle contents into the cells. Leserman et al.[273] have incubated SUV bearing the DNP-hapten on the surface and the fluorescent dye 6-carboxy-fluorescein in the aqueous phase with mouse myeloma tumour cells which secrete and possess on their surface IgA with affinity for the nitrophenyl hapten. They showed that, whereas good specific binding of SUV to the tumour cells occurred, there was no delivery of the vesicle contents (the dye) into the cytoplasm of the tumour cells. In this example, targetting (addressing and delivery) is achieved, but "entry" and "explosion" are lacking and so the system would fail as a drug delivery system to the tumour cells.

Liposomes are not the only carrier system under consideration for drug delivery; for example, DNA,[274] lectins,[275] glycoproteins,[276] and dextrans[277] all offer targetting possibilities. Nevertheless, the multi-step process will also be applicable to these other targetting molecules in order to achieve success. With our increasing knowledge of the properties of receptors on cells it may become possible to tailor-make the liposomes to fit the receptors on the target, and although this would still not guarantee delivery of contents into the cell it would surely be a step forward, and indeed the incorporation of specific glycolipids into liposomes has formed a useful model system for studying receptor–ligand interactions.[68]

One of the rather novel approaches which has recently been made to tumour targetting using liposomes involves the local release of drug from liposomes by hyperthermia induced at the target site. Yatvin et al.[278] have shown that there is a 100-fold enhancement of cell-kill when neomycin-containing dipalmitoyl-phosphatidylcholine liposomes (transition temperature (TT) approximately 41°C) are used to kill cells at 44°C compared with 37°C. This led the authors to suggest that hyperthermia of the tumour target might aid the delivery of anti-tumour agent if the phospholipids used in the preparation had a TT near that of the hyperthermic area. Weinstein and his colleagues[279] have obtained encouraging results in which the administration of methotrexate-containing vesicles led to a 4·5-fold greater accumulation of drug in the hyperthermic side of a double-tumoured animal in which one side was heated to 41°C.

Returning to the analogy of the letter bomb, the entry implies the crossing of the target cells membrane(s). Like many toxins, the lectins ricin, abrin and modeccin bind to cell surface receptors, enter the cell and therein exert their toxic effects. All these molecules appear to have A and B polypeptide components—the latter involved in binding to the cell surface and the former exerting its toxic effect on protein synthesis by binding to ribosomes. Diphtheria and cholera toxins appear to behave in a similar manner. The possibility of using such molecules either alone or in collaboration with liposomes to confer entry properties (B subunit) or killer potential (A subunit) is clearly very attractive. Hybrid toxin molecules, e.g. diphtheria A/concanavalin A,[280] ricin A/concanavalin A,[281] *Wisteria floribunda* lectin/diphtheria toxin A,[282] ricin A/human gonadotrophin[283] and diphtheria toxin A/anti-lymphocyte globulin[284] have all been prepared. This area seems of particular interest in tumour targetting when cell-kill is required and such hybrids seem to offer recognition and entry. Furthermore, the killer potential is very high as only a few molecules of toxin are necessary for cell death. Dimitriadis & Butters[285] and Gardas & MacPherson[286] have already begun work in the liposome direction and have shown liposome-mediated ricin toxicity in ricin-resistant cells.

Liposomes in all their various forms are not the only "vesicles" to be considered as possible carriers of therapeutic agents. Erythrocytes,[287, 288] polymorphonuclear leucocytes[289] and other white cells[290] have all to be considered, as have synthetic vesicles made of nylon,[291] lactic glycolic polymers,[292] albumin,[293] or starch microspheres.[294] Some of these carriers can be rendered magnetic by the inclusion of magnetite in them and perhaps could provide "magnetic" targetting.[293, 294] A novel colloidal drug delivery system in which nanoparticles of protein of 200 nm size can be prepared has been described by Speiser and his colleagues.[295]

There seems little doubt that the liposome as a carrier is a true "bag of potential" but much remains to be done before it can be clearly stated whether the promise that the system holds in a variety of fields will be realized. Whatever the outcome, the liposome field is an exciting one in which to work and it is full of interest.

ACKNOWLEDGEMENTS

The authors are indebted to the Medical and Science Research Councils, the Wellcome Trust and the Cancer Research Campaign for providing funds which have enabled us to work in the liposome field. Miss Linda Readings has provided unrivalled skills in the typing and preparation of the essay, and we thank her for these and for her ever-present good humour in the face of an avalanche of work. We are also grateful to Professor R. D. Marshall and Miss Christina Farren for their help in shortening our original Essay.

REFERENCES

1. Bangham, A. D. (1972). Lipid bilayers and biomembranes. *Annu. Rev. Biochem.* **41**, 753–776.
2. Bangham, A. D., Standish, M. M. & Watkins, J. C. (1965). The action of steroids and streptolysin S on the permeability of phospholipid structures to cations. *J. Mol. Biol.* **13**, 238–252.
3. Kornberg, R. D. & McConnell, H. M. (1971). Inside–outside transitions of phospholipids in vesicle membranes. *Biochemistry* **10**, 1111–1120.
4. Jost, P. C., Griffith, O. H., Capaldi, R. A. & Vanderkooi, G. (1973). Evidence for boundary lipid in membranes. *Proc. Natl. Acad. Sci. U.S.A.* **70**, 480–484.
5. Chapman, D. & Penkett, S. A. (1966). Effect of oestrogen on the small pigmented spots in hamsters. *Nature (London)* **211**, 1304–1305.
6. Stockton, G. W., Johnson, K. G., Butler, K. W., Polnaszek, C. F., Cyr, R. & Smith, I. C. P. (1975). Molecular order in *Acholeplasma laidlawii* membranes as determined by deuterium magnetic resonance of biosynthetically-incorporated specifically-labelled lipids. *Biochim. Biophys. Acta* **401**, 535–539.
7. Papahadjopoulos, D. (1972). Studies on the mechanism of action of local anaesthetics with phospholipid model membranes. *Biochim. Biophys. Acta* **265**, 169–186.
8. Tillack, T. W. & Kinsky, S. C. (1973). A freeze etch study of the effects of filipin on liposomes and human erythrocyte membranes. *Biochim. Biophys. Acta* **323**, 43–54.
9. Kinsky, S. C. & Nicolotti, R. A. (1977). Immunological properties of model membranes. *Annu. Rev. Biochem.* **46**, 49–68.
10. Alving, C. R. (1977). Immune reaction of lipids and lipid model membranes. In *Antigens* (Sela, M., ed.), pp. 1–72. Academic Press, New York and London.
11. Kimelberg, H. K. (1976). Protein liposome interactions and their relevance to the structure and function of cell membranes. *Mol. Cell. Biochem.* **10**, 171–190.
12. Gregoriadis, G., Leathwood, P. D. & Ryman, B. E. (1971). Enzyme entrapment in liposomes. *FEBS Lett.* **14**, 95–99.
13. Poste, G., Papahadjopoulos, D. & Vail, W. J. (1976). Lipid vesicles as carriers for introducing biologically active materials into cells. *Methods Cell Biol.* **14**, 33–71.
14. Tyrrell, D. A., Heath, T. D., Colley, C. M. & Ryman, B. E. (1976). New aspects of liposomes. *Biochim. Biophys. Acta* **457**, 259–302.
15. Papahadjopoulos, D. (ed.) (1978). Liposomes and their uses in biology and medicine. *Ann. N. Y. Acad. Sci.* **308**, 1–462.
16. Anner, B. M., Lane, L. K., Schwartz, A. & Pitts, B. J. R. (1977). A reconstituted $Na^+ + K^+$ pump in liposomes containing purified $Na^+ + K^+$ ATPase from kidney medulla. *Biochim. Biophys. Acta* **467**, 340–345.
17. Eytan, G. D. & Broza, R. (1978). Role of charge and fluidity in the incorporation of cytochrome oxidase into liposomes. *J. Biol. Chem.* **253**, 3196–3202.
18. Turner, M. J. & Sanderson, A. R. (1978). The preparation of liposomes bearing human (HLA) transplantation antigens. *Biochem. J.* **171**, 505–508.
19. Huang, C. (1969). Studies on phosphatidylcholine vesicles. Formation and physical characteristics. *Biochemistry* **8**, 344–352.
20. Huang, C. & Thompson, T. E. (1974). Preparation of homogeneous, single walled phosphatidylcholine vesicles. *Methods Enzymol.* **32**, 485–489.

21. Barenholzt, Y., Gibber, D., Litman, B. J., Goll, J., Thompson, T. E. & Carlson, F. D. (1977). A simple method for the preparation of homogeneous phospholipid vesicles. *Biochemistry* **16**, 2806–2810.
22. Batzri, S. & Korn, E. D. (1973). Single bilayer liposomes prepared without sonication. *Biochim. Biophys. Acta* **298**, 1015–1019.
23. Barenholzt, Y., Amselen, S. & Lichtenberg, D. (1979). A new method for preparation of phospholipid vesicles (liposomes). *FEBS Lett.* **99**, 210–214.
24. Papahadjopoulos, D., Vail, W. J., Jacobson, K. & Poste, G. (1975). Cochleate lipid cylinders. Formation by fusion of unilamellar lipid vesicles. *Biochim. Biophys. Acta* **394**, 483–491.
25. Deamer, D. & Bangham, A. D. (1976). Large volume liposomes by an ether vaporization method. *Biochim. Biophys. Acta* **443**, 629–634.
26. Deamer, D. W., Hill, M. W. & Bangham, A. D. (1976). Proton flux across liposome membranes. *Biophys. J.* **16A**, 111.
27. Schieren, H., Rudolph, S., Finkelstein, M., Coleman, P. & Weissmann, G. (1978). Comparison of large unilamellar vesicles prepared by a petroleum ether vaporisation method with multilamellar vesicles: ESR diffusion and entrapment analyses. *Biochim. Biophys. Acta* **542**, 137–153.
28. Milsmann, M. H. W., Schweidener, R. A. & Weder, H. G. (1978). The preparation of large single bilayer liposomes by a fast and controlled dialysis. *Biochim. Biophys. Acta* **512**, 147–155.
29. Enoch, H. G. & Strittmatter, P. (1979). Formation and properties of 1000 Å diameter single bilayer phospholipid vesicles. *Proc. Natl. Acad. Sci. U.S.A.* **76**, 145–149.
30. Szoka, F. & Papahadjopoulos, D. (1978). Procedure for preparation of liposomes with large internal aqueous space and high capture by reverse phase evaporation. *Proc. Natl. Acad. Sci. U.S.A.* **75**, 4194–4198.
31. Ryman, B. E. & Tyrrell, D. A. (1979). Liposomes—Methodology and applications. In *Lysosomes in Biology and Pathology* (Dingle, J. T., Jacques, P. J. & Shaw, I. H., eds.), Vol. 6, pp. 549–547. Elsevier/North-Holland Biomedical Press B. V., Amsterdam.
32. Gregoriadis, G. (1975). Enzyme-carrier potential of liposomes in enzyme replacement therapy. *New Eng. J. Med.* **292**, 704–710, 765–770.
33. Deguchi, K. & Miro, J. (1978). Solution properties of long chain dialkyldimethylammonium salt. 1. Formation of vesicles by dioctadecyldimethylammonium chloride. *J. Colloid. Interface Sci.* **65**, 155–161.
34. Chabala, J. C. & Shen, T. Y. (1978). The preparation of 3-cholesteryl 6-(glycosylthio)hexyl ethers and their incorporation into liposomes. *Carbohydr. Res.* **67**, 55–63.
35. Gregoriadis, G. (1976). Enzyme entrapment in liposomes. *Methods Enzymol.* **44**, 218–227.
36. Tyrrell, D. A., Richardson, V. J. & Ryman, B. E. (1977). The effect of serum protein fractions on liposome–cell interactions in cultured cells and the perfused rat liver. *Biochim. Biophys. Acta* **497**, 469–480.
37. Gregoriadis, G., Davisson, P. J. & Scott, S. (1977). Binding of drugs to liposome-entrapped macromolecules prevents diffusion of drugs from liposomes *in vitro* and *in vivo*. *Biochem. Soc. Trans.* **5**, 1323–1326.
38. Bangham, A. D. & Horne, R. W. (1964). Negative staining of phospholipids and their structural modification by surface-active agents as observed in the electron microscope. *J. Mol. Biol.* **8**, 660–668.

39. Bangham, A. D., Standish, M. M., Watkins, J. C. & Weissmann, G. (1967). The diffusion of ions from a phospholipid model membrane system. *Protoplasma* **63**, 183–187.

40. de Gier, J., Mandersloot, J. G. & van Deenen, L. L. M. (1968). Lipid composition and permeability of liposomes. *Biochim. Biophys. Acta* **150**, 666–675.

41. de Gier, J., Mandersloot, J. G., Hupkes, J. V., McElhaney, R. N. & van Beek, W. P. (1971). On the mechanism of non-electrolyte permeation through lipid bilayers and through biomembranes. *Biochim. Biophys. Acta* **233**, 610–618.

42. Weissmann, G., Sessa, G. & Weissmann, S. (1966). The action of steroids and Triton X-100 upon phospholipid/cholesterol structures. *Biochem. Pharmacol.* **15**, 1537–1551.

43. Weissmann, G., Sessa, G. & Weissmann, S. (1965). Effect of steroids and Triton X-100 on glucose filled phospholipid/cholesterol structures. *Nature (London)* **208**, 649–651.

44. Hill, M. W. (1978). Interaction of lipid vesicles with anaesthetics. *Ann. N.Y. Acad. Sci.* **308**, 101–110.

45. Galla, H.-J. & Sackmann, E. (1975). Chemically induced lipid phase separation in mixed vesicles containing phosphatidic acid. An optical study. *J. Am. Chem. Soc.* **97**, 4114–4120.

46. Smith, I. C. P., Tulloch, A. P., Stockton, G. W., Schreier, S., Joyce, A., Butler, K. W., Boulanger, Y., Blackwell, B. & Bennett, L. G. (1978). Determination of membrane properties at the molecular level by carbon-13 and deuterium magnetic resonance. *Ann. N.Y. Acad. Sci.* **308**, 8–28.

47. Jacobson, K. & Papahadjopoulos, D. (1975). Phase transitions and phase separations in phospholipid membranes induced by changes in temperature, pH and concentration of bivalent cations. *Biochemistry* **14**, 152–165.

48. Chapman, D., Peel, W. E. & Quinn, P. J. (1978). The modulation of bilayer fluidity by polypeptides and homogeneous catalysts. *Ann. N.Y. Acad. Sci.* **301**, 67–84.

49. Wilschut, J. C., Regts, J. & Scherphof, G. (1976). Reactivation of β-hydroxybutyrate dehydrogenase by phosphatidylcholine/phosphatidylethanolamine mixtures: evidence for a temperature induced phase separation. *FEBS Lett.* **63**, 328–332.

50. Eytan, G. D. & Broza, R. (1978). Selective incorporation of cytochrome oxidase into small liposomes. *FEBS Lett.* **85**, 175–178.

51. Saraste, M. (1978). Association of *Pseudomonas* cytochrome oxidase with liposomes. *Biochim. Biophys. Acta* **507**, 17–25.

52. Reavill, C. A., Wooster, M. S. & Plummer, D. T. (1978). The interaction of purified acetylcholinesterase from pig brain with liposomes. *Biochem. J.* **173**, 851–856.

53. Madeira, V. M. S. (1977). Incorporation of urease into liposomes. *Biochim. Biophys. Acta* **499**, 202–211.

54. Kaduce, T. L., Schmidt, R. W. & Spector, A. A. (1978). Acylcoenzyme A: cholesterol acyltransferase activity: solubilization and reconstitution in liposomes. *Biochem. Biophys. Res. Commun.* **81**, 462–468.

55. Brulet, P. & McConnell, H. M. (1976). Protein–lipid interactions: glycophorin and dipalmitoylphosphatidylcholine. *Biochem. Biophys. Res. Commun.* **68**, 363–368.

56. Redwood, W. R., Jansons, V. K. & Patel, B. C. (1975). Lectin receptor interactions in liposomes. *Biochim. Biophys. Acta* **406**, 347–361.
57. Shchipakin, V., Chuchlova, E. & Evtodienko, Y. (1976). Reconstitution of mitochondrial H-transporting system in proteoliposomes. *Biochem. Biophys. Res. Commun.* **69**, 123–127.
58. Ferguson, D. R. & Burton, K. A. (1977). Reconstitution in phospholipid vesicles of a glucose transport system from pig small intestine. *Nature (London)* **265**, 639–642.
59. Kasahara, M. & Hinkle, P. C. (1977). Reconstitution and purification of the D-glucose transport from human erythrocytes. *J. Biol. Chem.* **252**, 7384–7390.
60. Kasahara, M. & Hinkle, P. C. (1976). Reconstitution of D-glucose transport catalysed by a protein fraction from human erythrocytes in sonicated liposomes. *Proc. Natl. Acad. Sci. U.S.A.* **73**, 396–400.
61. Kinne, R. & Faust, R. G. (1977). Incorporation of D-glucose, L-alanine and phosphate-transport systems from rat renal brush-border membranes into liposomes. *Biochem. J.* **168**, 311–314.
62. Ingelman-Sundberg, M. & Glaumann, H. (1977). Reconstitution of the liver microsomal hydroxylase system into liposomes. *FEBS Lett.* **78**, 72–76.
63. Hilden, S. & Hokin, L. (1976). Coupled $Na^+–K^+$ transport in vesicles containing a purified (NaK)-ATPase and only phosphatidylcholine. *Biochem. Biophys. Res. Commun.* **69**, 506–513.
64. Zilversmit, D. B. (1978). Phospholipid-exchange proteins as membrane probes. *Ann. N.Y. Acad. Sci.* **301**, 149–163.
65. Wirtz, K. W. A. & Zilversmit, D. B. (1968). Exchange of phospholipids between liver mitochondria and microsomes *in vitro. J. Biol. Chem.* **243**, 3596–3602.
66. Aloj, S. M., Kohn, L. D., Lee, G. & Meldolesi, M. F. (1977). Binding of thyrotropin to liposomes containing gangliosides. *Biochem. Biophys. Res. Commun.* **74**, 1053–1059.
67. Surolia, A., Bachhawat, B. K. & Podder, S. K. (1975). Interaction between lectin from *Ricinus communis* and liposomes containing gangliosides. *Nature (London)* **257**, 802–804.
68. Surolia, A. & Bacchawat, B. K. (1978). The effect of lipid composition on liposome–lectin interaction. *Biochem. Biophys. Res. Commun.* **83**, 779–785.
69. Maget-Dard, R., Roche, A. C. & Monsigny, M. (1977). Interactions between vesicles containing gangliosides and lectins, Limulin and wheat germ agglutinin. *FEBS Lett.* **79**, 305–309.
70. Littman, D. E., Cullen, S. E. & Schwartz, B. D. (1979). Insertion of Ia and H-2 alloantigens into model membranes. *Proc. Natl. Acad. Sci. U.S.A.* **76**, 902–906.
71. Rapport, M. M. & Graf, L. (1969). Immunochemical reactions of lipids. *Prog. Allergy* **13**, 273–331.
72. Kinsky, S. C. (1978). Immunogenicity of liposomal model membranes. *Ann. N.Y. Acad. Sci.* **301**, 111–123.
73. Uemura, K., Nicolotti, R. A., Six, H. R. & Kinsky, S. C. (1974). Antibody formation in response to liposomal model membranes sensitized with *N*-substituted phosphatidylethanolamine derivatives. *Biochemistry* **13**, 1572–1578.
74. Nicolotti, R. A. & Kinsky, S. C. (1975). Immunogenicity of liposomal model membranes sensitized with mono (*p*-azo-benzeneazonic acid)-tryosyl phosphatidyl ethanolamine derivatives. Antibody formation and delayed hypersensitivity reaction. *Biochemistry* **14**, 2331–2338.

75. Yasuda, T., Dancey G. F. & Kinsky, S. C. (1977). Immunogenic properties of liposomal model membranes in mice. *J. Immunol.* **119**, 1863–1867.
76. Dancey, G. F., Yasuda, T. & Kinsky, S. C. (1977). Enhancement of liposomal model membrane immunogenicity by incoporation of lipid A. *J. Immunol.* **119**, 1868–1873.
77. Dancey, G. F., Isakson, P. C. & Kinsky, S. C. (1979). Immunogenicity of liposomal model membranes sensitized with dinitrophenylated phosphatidylethanolamine derivatives containing different length spacers. *J. Immunol.* **122**, 638–642.
78. Schuster, B. G., Neidig, M., Alving, B. M. & Alving, C. R. (1979). Production of antibodies against phosphocholine, phosphatidylcholine, sphingomyelin and lipid A by injection of liposomes containing lipid A. *J. Immunol.* **122**, 900–905.
79. Yasuda, T., Dancey, G. F. & Kinsky, S. C. (1977). Immunogenicity of liposomal model membranes in mice: Dependence on phospholipid composition. *Proc. Natl. Acad. Sci. U.S.A.* **74**, 1234–1236.
80. Kinsky, S. C. (1972). Antibody-complement interaction with lipid model membranes. *Biochim. Biophys. Acta* **265**, 1–23.
81. Kinsky, S. C. (1974). Preparation of liposomes and a spectrophotometric assay for release of trapped glucose marker. *Methods Enzymol.* **32**, 501–513.
82. Kinoshita, T., Inoue, K., Okada, M. & Akiyama, Y. (1977). Release of phospholipids from liposomal model membrane damaged by antibody and complement. *J. Immunol.* **119**, 75–78.
83. Inoue, K. & Kinsky, S. C. (1970). Fate of phospholipids in liposomal model membranes damaged by antibody and complement. *Biochemistry* **9**, 4767–4776.
84. Chan, S. W., Tan, C. T. & Hsia, J. C. (1977). Antiliposome antisera activity against negatively-charged phosphate amphiphils. *Biochem. Biophys. Res. Commun.* **79**, 631–634.
85. Knudson, K. C., Bing, D. H. & Kater, L. (1971). Quantitative measurement of guinea pig complement with liposomes. *J. Immunol.* **106**, 258–265.
86. Conrad, D. H., Alving, C. R. & Wirtz, G. H. (1974). The influence of retinal on complement-dependent immune damage to liposomes. *Biochim. Biophys. Acta* **332**, 36–46.
87. Hsia, J. C. & Tan, C. T. (1978). Membrane immunoassay: Principle and applications of spin membrane immunoassay. *Ann. N.Y. Acad. Sci.* **308**, 139–148.
88. Lewis, J. T. & McConnell, H. M. (1978). Model lipid bilayer membranes as targets for antibody-dependent cellular and complement-mediated immune attack. *Ann. N.Y. Acad. Sci.* **308**, 124–138.
89. Esser, A. F., Bartholomew, R. M., Parce, J. W. & McConnell, H. M. (1979). The physical state of membrane lipids modulate the activation of the first component of complement. *J. Biol. Chem.* **254**, 1768–1770.
90. Richards, R. L., Gewurz, H., Osmand, A. P. & Alving, C. R. (1977). Interactions of C-reactive protein and complement with liposomes. *Proc. Natl. Acad. Sci. U.S.A.* **74**, 5672–5676.
91. Ozato, K., Ziegler, H. K. & Henney, C. S. (1978). Liposomes as model membrane systems for immune attack. *J. Immunol.* **121**, 1383–1388.
92. Cunningham, C. M., Kingzette, M., Richards, R. L., Alving, C. R., Lint, T. F. & Gewurz, H. (1979). Activation of human complement by liposomes: A model for membrane activation of the alternative pathway. *J. Immunol.* **122**, 1237–1242.
93. Heath, T. D., Edwards, D. C. & Ryman, B. E. (1976). The adjuvant properties of liposomes. *Biochem. Soc. Trans.* **4**, 129–132.

94. Allison, A. C. & Gregoriadis, G. (1974). Liposomes as immunological adjuvants. *Nature (London)* **252**, 252.
95. Allison, A. C. & Gregoriadis, G. (1976). Liposomes as immunological adjuvants. *Recent Results Cancer Res.* **56**, 58–64.
96. Haywood, A. M. (1978). Interaction of liposomes with viruses. *Ann. N.Y. Acad. Sci.* **308**, 275–280.
97. Barratt, D. G., Sharon, F. J., Thede, A. E. & Gant, C. W. M. (1977). Isolation and incorporation into lipid vesicles of a concanavalin A receptor from human erythrocytes. *Biochim. Biophys. Acta* **465**, 191–197.
98. Pagano, R. E. & Weinstein, J. N. (1978). Interactions of liposomes with mammalian cells. *Annu. Rev. Biophys. Bioeng.* **7**, 435–468.
99. Poste, G. & Papahadjopoulos, D. (1978). The influence of vesicle membrane properties on the interaction of lipid vesicles with cultured cells. *Ann. N.Y. Acad. Sci.* **308**, 164–185.
100. Papahadjopoulos, D., Poste, G. & Mayhew, E. (1974). Cellular uptake of cyclic AMP captured within phospholipid vesicles and effect on cell growth behaviour. *Biochim. Biophys. Acta* **363**, 404–418.
101. Cohen, C. M., Weissmann, G., Hoffstein, S., Awarthi, Y. C. & Srivastava, S. K. (1976). Introduction of purified hexosaminidase A into Tay-Sachs leukocytes by means of immunoglobulin-coated liposomes. *Biochemistry* **15**, 452–460.
102. Martin, F. J. & MacDonald, R. C. (1976). Lipid vesicle–cell interactions. *J. Cell Biol.* **70**, 494–505.
103. Huang, L., Ozato, K. & Pagano, R. E. (1978). Interactions of phospholipid vesicles with murine lymphocytes. I. Vesicle-cell adsorption and fusion as alternative pathways of uptake. *Membrane Biochem.* **1**, 1–25.
104. Huang, L. & Pagano, R. E. (1975). Interaction of phospholipid vesicles with cultured mammalian cells. I. Characteristics of uptake. *J. Cell Biol.* **67**, 38–48.
105. Pagano, R. E. & Huang, L. (1975). Interaction of phospholipid vesicles with cultured mammalian cells. II. Studies of mechanism. *J. Cell Biol.* **67**, 49–60.
106. Pagano, R. E. & Takeichi, M. (1977). Adhesion of phospholipid vesicles to Chinese hamster fibroblasts. Role of cell surface proteins. *J. Cell Biol.* **74**, 531–546.
107. Sandra, A. & Pagano, R. E. (1979). Liposome–cell interactions. Studies of lipid transfer using isotopically asymmetric vesicles. *J. Biol. Chem.* **254**, 2244–2249.
108. Poste, G. & Papahadjopoulos, D. (1976). Lipid vesicles as carriers for introducing materials into cultured cells—influence of vesicle lipid composition on mechanism(s) of vesicle incorporation into cells. *Proc. Natl. Acad. Sci. U.S.A.* **73**, 1603–1607.
109. Batzri, S. & Korn, E. D. (1975). Interaction of phospholipid vesicles with cells. Endocytosis and fusion as alternate mechanisms for the uptake of lipid soluble and water soluble molecules. *J. Cell Biol.* **66**, 621–635.
110. Weissmann, G., Bloomgarden, D., Kaplan, R., Cohen, C., Hoffstein, S., Collins, T., Gottlieb, A. & Nagle, D. (1975). A general method for the introduction of enzymes by means of immunoglobulin coated liposomes into lysosomes of deficient cells. *Proc. Natl. Acad. Sci. U. S.A.* **72**, 88–92.
111. Weissmann, G., Cohen, C. & Hoffstein, S. (1977). Introduction of enzymes by means of liposomes jinto non-phagocytic human cells *in vitro*. *Biochim. Biophys. Acta* **498**, 375–385.
112. Martin, F. J. & MacDonald, R. C. (1976). Lipid vesicle cell interations. II. Introduction of cell fusion. *J. Cell Biol.* **70**, 506–514.

113. Papahadjopoulos, D., Poste, G. & Mayhew, E. (1975). The interaction of phospholipid vesicles with mammalian cells *in vitro*. *Biochem. Soc. Trans.* **3**, 606–608.

114. Weinstein, J. N., Yoshikami, S., Henkart, P., Blumenthal, R. & Hagins, W. A. (1977). Liposome-cell interaction. Transfer and intracellular release of a trapped fluorescent marker. *Science* **195**, 489–492.

115. Blumenthal, R., Weinstein, J. N., Harrow, S. O. & Henkart, P. (1977). Liposome–lymphocyte interaction: saturable sites for transfer and intracellular release of liposome contents. *Proc. Natl. Acad. Sci. U.S.A.* **74**, 5603–5607.

116. Szoka, F. C., Jacobson, K. & Papahadjopoulos, D. (1979). The use of aqueous space markers to determine the mechanism of interaction between phospholipid vesicles and cells. *Biochim. Biophys. Acta* **551**, 295–303.

117. Gregoriadis, G. & Buckland, R. A. (1973). Enzyme-containing liposomes alleviate a model for storage disease. *Nature* **244**, 170–172.

118. Roerdink, F. H., van Renswoude, A. J. B. M., Wielinga, B. Y., Kroon, A. M. & Scherphof, G. L. (1976). Entrapment within liposomes facilitates uptake of amyloglucosidase by cultured fibroblasts from a patient with Pompe's disease (glycogenosis type II). *J. Mol. Med.* **1**, 257–264.

119. Poste, G. & Papahadjopoulos, D. (1976). Drug-containing lipid vesicles render drug resistant tumour cells sensitive to actinomycin D. *Nature (London)* **261**, 699–701.

120. Ostro, M. J., Giacomoni, D. & Dray, S. (1977). Incorporation of high molecular weight RNA into large artificial lipid vesicles. *Biochem. Biophys. Res. Commun.* **76**, 836–842.

121. Hoffman, R. M., Margolis, L. B. & Bergelson, L. D. (1978). Binding and entrapment of high molecular weight DNA by lecithin liposomes. *FEBS Lett.* **93**, 365–368.

122. Ostro, M. J., Giacomoni, D., Lavelle, D., Paxton, W. & Dray, S. (1978). Evidence for translation of rabbit globin mRNA after liposome mediated insertion into a human cell line. *Nature (London)* **274**, 921–923.

123. Dimiatriadis, G. J. (1978). Translation of rabbit globin mRNA introduced by liposomes into mouse lymphocytes. *Nature (London)* **274**, 923–924.

124. Gregoriadis, G. & Neerunjun, E. D. (1975). Homing of liposomes to target cells. *Biochem. Biophys. Res. Commun.* **65**, 537–544.

125. Bussian, R. W. & Writon, J. C. (1977). Influence of incorporated cerebrosides on the interaction of liposomes with HeLa cells. *Biochim. Biophys. Acta* **471**, 336–340.

126. Juliano, R. L. & Stamp, D. (1976). Lectin-mediated attachment of glycoprotein-bearing liposomes to cells. *Nature (London)* **261**, 235–237.

127. Margolis, L. B. & Dorfman, N. A. (1977). Preparation of liposomes with immunological specificity. *Bull. Exp. Biol. Med.* **83**, 60.

128. Klareskog, L., Banck, G., Forsgren, A. & Peterson, P. A. (1978). Binding of HLA antigen-containing liposomes to bacteria. *Proc. Natl. Acad. Sci. U.S.A.* **75**, 6197–6201.

129. Schroit, A. J. & Pagano, R. E. (1978). Introduction of antigenic phospholipids into the plasma membrane of mammalian cells: organisation and antibody-induced lipid redistribution. *Proc. Natl. Acad. Sci. U.S.A.* **75**, 5529–5533.

130. Gregoriadis, G. & Ryman, B. E. (1972). Fate of protein-containing liposomes injected into rats. An approach to the treatment of storage diseases. *Eur. J. Biochem.* **24**, 485–491.

131. Gregoriadis, G. & Ryman, B. E. (1972). Lysosomal localisation of β-fructofuranosidase containing liposomes injected into rats—implications in the treatment of genetic disorders. *Biochem. J.* **129**, 123–133.
132. Juliano, R. L. & Stamp, D. (1975). The effect of particle size and charge on the clearance rate of liposomes and liposome encapsulated drug. *Biochem. Biophys. Res. Commun.* **63**, 651–658.
133. Sharma, P., Tyrrell, D. A. & Ryman, B. E. (1977). Some properties of liposomes of different sizes. *Biochem. Soc. Trans.* **5**, 1146–1149.
134. Wharton, S. A. & Green, C. (1978). Plasma clearance and tissue distribution of liposomes containing different sterols. *Biochem. Soc. Trans.* **6**, 781–783.
135. Segal, A. W., Gregoriadis, G. & Black, C. D. V. (1975). Liposomes as vehicles for the local release of drugs. *Clin. Sci. Mol. Med.* **49**, 99–106.
136. Tanaka, T., Taneda, K., Kobayashi, H., Okumura, K., Maranishi, S. & Sezaki, H. (1975). Application of liposomes to the pharmaceutical modification of the distribution characteristics of drugs in the rat. *Chem. Pharm. Bull.* **23**, 3069–3074.
137. Ryman, B. E., Jewkes, R. F., Jeyasingh, K., Osborne, M. P., Patel, H. M., Richardson, V. J., Tattersall, M. H. N. & Tyrrell, D. A. (1978). Potential applications of liposomes to therapy. *Ann. N.Y. Acad. Sci.* **308**, 281–307.
138. Rustum, Y. M., Dave, C., Meyhew, E. & Papahadjopoulos, D. (1979). Role of liposome type and route of administration in the antitumor activity of liposome-entrapped 1-β-D-arabinofuranosyl cytosine against mouse L1210 leukaemia. *Cancer Res.* **39**, 1390–1395.
139. Patel, H. M. & Ryman, B. E. (1976). Oral administration of insulin by encapsulation within liposomes. *FEBS Lett.* **62**, 60–64.
140. Gregoriadis, G., Dapergolas, G. & Neerunjun, E. D. (1976). Penetration of target areas in the rat by liposome-associated agents administered parenterally and intragastrically. *Biochem. Soc. Trans.* **4**, 256–258.
141. Gregoriadis, G. (1977). Targetting of drugs. *Nature (London)* **265**, 407–411.
142. Gregoriadis, G. (1976). The carrier potential of liposomes in biology and medicine. *New Eng. J. Med.* **295**, 704–710, 765–770.
143. Gregoriadis, G. (ed.) (1979). *Drug Carriers in Biology and Medicine.* Academic Press, London and New York.
144. Sessa, G. & Weissman, G. J. (1969). Formation of artificial lysosome *in vitro. J. Clin. Invest.* **48**, 76a–77a.
145. Reynolds, G. D., Baker, H. J. & Reynolds, R. H. (1978). Enzyme. replacement using liposome carriers in feline G_{ml} gangliosidosis fibroblasts. *Nature (London)* **275**, 754–755.
146. Butler, J. de B., Tietze, F., Pelletigue, F., Spielberg, S. P. & Schulman, J. D. (1978). Depletion of cystine in cystinotic fibroblasts by drugs enclosed in liposomes. *Pediatr. Res.* **12**, 46–51.
147. Millard, P. C. (1979). Studies on the release of exogenous molecules from rat liver lysosomes. *Ph.D. Thesis*, University of Keele.
148. Heath, T. D. (1976). The effect of liposomes upon the immune response. *Ph.D. Thesis*, University of London.
149. Gregoriadis, G. & Allison, A. C. (1974). Entrapment of protein in liposomes prevents allergic reactions in pre-immunised mice. *FEBS Lett.* **45**, 71–74.
150. Hudson, L. D. S., Fiddler, M. B. & Desnick, R. J. (1979). Enzyme therapy. X. Immune response induced by enzyme- and buffer-loaded liposomes in C3H/HeJ Gus[h] mice. *J. Pharmacol. Exp. Ther.* **208**, 507–514.

151. Tyrrell, D. A., Ryman, B. E., Keeton, B. R. & Dubowitz, V. (1976). Use of liposomes in treating Type II glycogenosis. *Br. Med. J.* **ii**, 88–89.
152. Belchetz, P. E., Braidman, I. P., Crawley, J. C. W. & Gregoriadis, G. (1971). Treatment of Gaucher's disease with liposome-entrapped glucocerebroside: β-glucosidase. *Lancet* **i**, 116–117.
153. Finkelstein, M. & Weissmann, G. (1979). The introduction of enzymes into cells by means of liposomes. *J. Lipid Res.* **19**, 289–303.
154. Finkelstein, M. & Weissmann, G. (1979). Enzyme replacement via liposomes: variations in lipid composition determine liposomal integrity in biological fluids. *Biochim. Biophys. Acta* **587**, 202–216.
155. Weissmann, G., Korchak, H., Finkelstein, M., Smolen, J. & Hoffstein, S. (1978). Uptake of enzyme-laden liposomes by animals cells *in vitro* and *in vivo*. *Ann. N.Y. Acad. Sci.* **308**, 235–249.
156. Gregoriadis, G. (1979). Liposomes. In *Drug Carriers in Biology and Medicine* (Gregoriadis, G., ed.), pp. 287–341. Academic Press, London and New York.
157. Trouet, A. (1978). Perspectives in cancer research. Increased selectivity of drugs by linking to carriers. *Eur. J. Cancer* **14**, 105–111.
158. Papahadjopoulos, D., Poste, G., Vail, W. J. & Biedler, J. L. (1976). Use of lipid vesicles as carriers to introduce actinomycin D into resistant tumour cells. *Cancer Res.* **36**, 2988–2994.
159. Mayhew, E., Papahadjopoulos, D., Rustum, Y. M. & Dave, C. (1978). Use of liposomes for the enhancement of the cytotoxic effects of cytosine arabinoside. *Ann. N.Y. Acad. Sci.* **308**, 371–384.
160. Mayhew, E., Rustum, Y. M., Szoka, F. & Papahadjopoulos, D. (1979). Role of cholesterol in enhancing the antitumor activity of cytosine arabinoside entrapped in liposomes. *Cancer Treat. Rep.* **63**, 1928.
161. Juliano, R. L. & Stamp, D. (1978). Pharmacokinetics of liposome-encapsulated anti-tumor drugs. Studies with vinblastine, actinomycin D, cytosine arabinoside and daunomycin. *Biochem. Pharmacol.* **27**, 21–27.
162. Kaye, S. B. (1979). An evaluation of liposomes as drug carriers in cancer chemotherapy. *M.D. Thesis*, University of London.
163. Kaye, S. B. & Ryman, B. E. (1980). The *in vivo* fate of liposome-entrapped actinomycin D and its therapeutic effect in a solid murine tumour. *Biochem. Soc. Trans.* **8**, 107–108.
164. Kaye, S. B., Boden, J. A. & Ryman, B. E. (1980). The effect of liposome (phospholipid vesicle) entrapment of actinomycin D and methotrexate on the *in vivo* treatment of sensitive and resistant solid murine tumours. *Eur. J. Cancer*, in press.
165. Kosloski, M. J., Rosen, F., Milholland, R. J. & Papahadjopoulos, D. (1978). Effect of lipid vesicle (liposome) encapsulation of methotrexate on its chemotherapeutic efficacy in solid rodent tumors. *Cancer Res.* **38**, 2848–2853.
166. Kimelberg, H. K., Tracy, T. F., Watson, R. E., Kung, D., Reiss, F. L. & Bourke, R. S. (1978). Distribution of free and liposome-entrapped [³H]methotrexate in the central nervous system after intracerebroventricular injection in a primate. *Cancer Res.* **38**, 706–712.
167. Adams, D. H., Joyce, G., Richardson, V. J., Ryman, B. E. & Wisniewski, H. M. (1977). Liposome-toxicity in the mouse central nervous system. *J. Neur. Sci.* **31**, 173–179.
168. McCullough, H. N. & Juliano, R. L. (1979). Organ-selective action of anti-tumor drug—pharmacological study of liposome entrapped beta-cytosine arabinoside

administered via the respiratory system of the rat. *J. Natl. Cancer Inst.* **63**, 727–731.

169. Hunt, C. A., Rustum, Y. M., Mayhew, E. & Papahadjopoulos, D. (1979). Retention of cytosine arabinoside in mouse lung following intravenous administration in liposomes of different size. *Drug Metab. Dispos.* **17**, 124–128.

170. Richardson, V. J., Jeyasingh, K., Jewkes, R. F., Ryman, B. E. & Tattersall, M. H. N. (1978). Possible localisation of ^{99m}Tc-labelled liposomes: effects of lipid composition, charge and liposome size. *J. Nucl. Med.* **19**, 1049–1054.

171. Angileri, L. J., Firusian, N. & Brucksch, K. P. (1976). *In vivo* distribution of ^{99m}-technetium labelled liposome. *J. Nucl. Biol. Med.* **20**, 165–167.

172. Gregoriadis, G., Swain, C. P., Wills, E. J. & Tavill, A. S. (1974). Drug-carrier potential of liposomes in cancer chemotherapy. *Lancet* i, 1313–1316.

173. Richardson, V. J., Ryman, B. E., Jewkes, R. F., Jeyasingh, K., Tattersall, M. H. N., Newlands, E. S. & Kaye, S. B. (1979). Tissue distribution and tumour localization of 99m-technetium-labelled liposomes in cancer patients. *Br. J. Cancer* **40**, 35–43.

174. Richardson, V. J., Ryman, B. E., Jewkes, R. F., Tattersall, M. H. N. & Newlands, E. S. (1978). ^{99m}Tc-Labelled liposomes preparation of radiopharmaceutical and its distribution in a hepatoma patient. *Int. J. Nucl. Med. Biol.* **5**, 118–123.

175. Magee, W. C., Cronenberger, J. H. & Thor, D. E. (1978). Marked stimulation of lymphocyte-mediated attack on tumor cells by target-directed liposomes containing immune RNA. *Cancer Res.* **38**, 1173–1176.

176. Osborne, M. P., Richardson, V. J., Jeyasingh, K. & Ryman, B. E. (1979). Radionuclide-labelled liposomes—a new lymph node imaging agent. *Int. J. Nucl. Med. Biol.* **6**, 75–83.

177. Richardson, V. J., Osborne, M. P., Jeyasingh, K., Ryman, B. E. & Burn, J. I. (1978). Differential localisation of [^{99m}Tc]technetium-labelled liposomes in normal and tumour-bearing lymph nodes of the rat. *Br. J. Cancer* **38**, 177.

178. Osborne, M. P., Richardson, V. J., Payne, J. H., McCready, V. R. & Ryman, B. E. (1979). Technetium-99m labelled liposome axillary lymphoscintigraphy in breast cancer. BASO Meeting, University College Hospital, London, December 7, 1979.

179. Rahman, Y. E. (1979). Potential of the liposomal approach to metal chelation therapy. In *Lysosomes in Biology and Pathology* (Dingle, J. T., Jacques, P. J. & Shaw, I. H., eds.), Vol. 6, pp. 625–652. Elsevier-North-Holland Biomedical Press, Amsterdam.

180. Jonah, M. M., Cerny, E. A. & Rahman, Y. E. (1978). Tissue distribution of EDTA encapsulated within liposomes containing glycolipids or brain phospholipids. *Biochim. Biophys. Acta* **541**, 321–333.

181. Ryman, B. E. & Seymour, A.-M. L. (1978). Substitution therapy in lysosomal storage diseases. *12th FEBS Meeting, Dresden, 1978*, **56**, *Colloquium G3*, Molecular Diseases (Schewe, T. & Rapoport, S., eds.), pp. 79–88. Pergamon Press, Oxford.

182. Seymour, A.-M. L. (1980). Drug entrapment in liposomes. Some biochemical and chemical aspects. Ph.D. Thesis, University of London.

183. Young, S. P., Baker, E. & Huens, E. R. (1979). Liposome entrapped desferrioxamine and iron transporting ionophores: A new approach to iron chelation therapy. *Br. J. Haematol.* **41**, 357–363.

184. Dingle, J. T., Gordon, J. L., Hazleman, B. E., Knight, C. G., Page Thomas, D. P., Phillips, N. C., Shaw, I. H., Fildes, F. J. T., Oliver, J. E., Jones, G., Turner, E. H. & Lowe, J. S. (1978). Novel treatment for joint inflammation. *Nature* (*London*) **271**, 372–373.

185. Shaw, I. H., Knight, C. G., Page Thomas, D. P. & Phillips, N. C. (1979). Liposome-incorporated corticosteroids: I. The interaction of liposomal cortisol palmitate with inflammatory synovial membrane. *Br. J. Exp. Pathol.* **60**, 142–150.

186. Shaw, I. H., Knight, C. G. & Dingle, J. T. (1976). Liposomal retention of a modified anti-inflammatory steroid. *Biochem. J.* **158**, 473–476.

187. de Silva, M., Page Thomas, D. P., Hazleman, B. L. & Wraight, P. (1979). Liposomes in arthritis: A new approach *Lancet* **i**, 1320–1322

188. Finkelstein, M. C., Maniscalo, J. & Weissmann, G. (1978). Entrapment of soy bean trypsin inhibitor and α_1-antitrypsin by multilamellar liposomes. *Anal. Biochem.* **89**, 400–407.

189. Tragl, K. H., Pohl, A. & Kinast, H. (1979). Zur perotalen verabreichung voninsulin millels liposomen im Tierversuch. *Wien. Klin. Wochenschr.* **91**, 448–451.

190. Hashimoto, A. & Kawada, J. (1979). Effects of oral administration of positively-charged insulin-liposomes on alloxan diabetic rats. *Endocrinology* (*Japan*) **26**, 337–343.

191. Patel, H. M., Harding, N. G. L., Logue, F., Kesson, C., MacCuish, A. C., McKenzie, J. C., Ryman, B. E. & Scobie, I. (1978). Intrajejunal absorption of liposomally-entrapped insulin in normal man. *Biochem. Soc. Trans.* **6**, 784–785:

192. Bridges, J. F., Millard, P. C. & Woodley, J. F. (1978). The uptake of liposome-entrapped ^{125}I-labelled poly(vinylpyrrolidone) by rate jejunum *in vitro*. *Biochim. Biophys. Acta* **544**, 448–451.

193. Whitmore, D. A. & Wheeler, K. P. (1980). Partial purification of a glucose-transport system from rat jejunum with the use of a cationic surfactant. *Biochem. Soc. Trans.*, **8**, 318.

194. Hori, R., Kagimoto, Y. & Inui, K. I. (1978). Effects of free fatty acids as membrane components on permeability of drugs across bilayer lipid membranes: a mechanism for intestinal absorption of acidic drugs. *Biochim. Biophys. Acta* **509**, 510–518.

195. Richards, M. H. & Gardner, C. R. (1978). Effects of bile salts on the structural integrity of liposomes. *Biochim. Biophys. Acta* **543**, 508–522.

196. Hirai, S., Ikenaga, T. & Matsuzawa, T. (1978). Nasal absorption of insulin in dogs. *Diabetes* **27**, 296–299.

197. Theurer, K. Orally active drugs and vaccines from primary non-intestinally active preparations. *Ger. Offen.* 2 645 444 (Cl.A61K9/10) 13 Apr. 1978. Appl. 08 Oct 1976.

198. Theurer, K. Protein and peptide solutions resorbable in the intestine. *Ger. Offen.* 2 640 707 (Cl.A61K37/02). 16 Mar. 1978. App. 10 Sep. 1976.

200. Stevens, R. W., Patel, H. M., Parsons, J. A. & Ryman, B. E. (1980). Subcutaneous administration of liposomes containing insulin to alloxan-streptozotocin diabetic dogs. *J. Clin. Invest.*, in press.

201. Black, C. D. V., Watson, G. J. & Ward, R. W. (1977). The use of pentostam liposomes in the chemotherapy of experimental leishmaniasis. *Trans. R. Soc. Trop. Med. Hyg.* **71**, 550–552.

202. New, R. R. C., Chance, M. L., Thomas, S. C. & Peters, W. (1978). Antileishmaniasis activity of antimonials entrapped in liposomes. *Nature (London)* **272**, 55–56.

203. Alving, C. R., Steck, E. A., Chapman, W. L., Waits, V B., Hendricks, L. D., Swartz, G. M. & Hanson, W. L. (1978). Therapy of leishmaniasis: superior efficacies of liposome-encapsulated drugs. *Proc. Natl. Acad. Sci. U.S.A.* **75**, 2959–2953.

204. Alving, C. R., Schneider, I., Swartz, G. M., Jr. & Steck, E. A. (1979). Sporozoite-induced malaria: therapeutic effects of glycolipids in liposomes. *Science* **205**, 1142–1144.

205. Alving, C. R. & Steck, E. A. (1979). The use of liposome-encapsulated drugs in leishmaniasis. *Trends Biochem. Sci.* **4**, N175–N177.

206. Ashwell, G. & Morell, A. G. (1977). Membrane glycoproteins and recognition phenomena. *Trends Biochem. Sci.* **2**, 76–78.

207. Vakirtzi-Lemonias, C. & Gregoriadis, G. (1978). Uptake of liposome-entrapped agents by the Trypanosome: *Crithidia fasculata. Biochem. Soc. Trans.* **6**, 1241–1244.

208. Gruenburg, J., Coral, D., Knupfer, A. L. & Deshusses, J. (1979). Interactions of liposomes with *Trypanosoma brucei* plasma membrane. *Biochem. Biophys. Res. Commun.* **88**, 1173–1179.

209. Morley, C. J., Bangham, A. D., Johnson, P., Thorburn, G. D. & Jenkin, G. (1978). Physical and physiological properties of dry lung surfactant. *Nature (London)* **271**, 162–163.

210. Blumenfeld, T. A., Stark, R. I., James, L. S., George, J. D., Dyrenfurth, I., Freda, V. J. & Shinitzky, M. (1978). Determination of fetal lung maturity by fluorescence polarization of amniotic fluid. *Am. J. Obstet. Gynecol.* **130**, 782–787.

211. Tyrrell, D. A., Campbell, P. I., Harding, N. G. L., Munro, A. & Ryman, B. E. (1978). Antidigoxin antibody incorporation into liposomes—a potential therapy for digoxin toxicity. *Biochem. Soc. Trans.* **6**, 1239–1241.

212. Campbell, P. I., Harding, N. G. L., Ryman, B. E. & Tyrrell, D. A. (1980) Redistribution and altered excretion of digoxin in rats receiving digoxin antibodies incorporated in liposomes. *Eur. J. Biochem.*, in press.

213. Caride, V. J. & Zaret, B. L. (1977). Liposome accumulation in regions of experimental myocardial infarction. *Science* **198**, 735–738.

214. Caride, V. J., Taylor, W., Cramer, J. A. & Gottschalk, A. (1976). Evaluation of liposome-entrapped radioactive tracers as scanning agents. I. Organ distribution of liposome (99mTc-DTPA) in mice. *J. Nucl. Med.* **17**, 1067–1072.

215. Espinola, L. G., Beaucaire, J., Gottschalk, A. & Caride, V. J. (1979). Radiolabelled liposomes as metabolic and scanning tracers in mice. II. In-111 compared with Tc-99m DTPA, entrapped in multilamellar lipid vesicles. *J. Nucl. Med.* **20**, 434–440.

216. van Rooijen, N. & van Nieuwmeger, R. (1978). Liposomes in immunology: further evidence for the adjuvant activity of liposomes. *Immunol. Commun.* **7**, 635–644; (1980) *Cell. Immunol.* **49**, 402–407.

217. van Rooijen, N. & van Nieuwmeger, R. (1979). Liposomes in immunology: impairment of the adjuvant effect of liposomes by incorporation of the adjuvant lysolecithin and the role of macrophages. *Immunol. Commun.* **8**, 381–396.

218. Haywood, A. M. (1975). "Phagocytosis" of Sendai virus by model membranes. *J. Gen. Virol.* **29**, 63–68.

219. Almeida, J. D., Edwards, D. C., Brand, C. M. & Heath, T. D. (1975). Formation of virosomes from influenza subunits and liposomes. *Lancet* **ii**, p. 899.
220. Morein, B., Helenius, A., Simons, K., Pettersson, R., Kääriäiner, L. & Schirrmacher, V. (1978). Effective subunit vaccines against an enveloped animal virus. *Nature (London)* **276**, 715–718.
221. Manesis, E. K., Cameron, C. H. & Gregoriadis, G. (1979). Hepatitis-B-surface-antigen-containing liposomes enhance humoral and cell-mediated immunity to the antigen. *Biochem. Soc. Trans.* **7**, 678–680.
222. Kramp, W. J., Six, H. R., Drake, S. & Kasel, J. A. (1979). Liposomal enhancement of the immunogenicity of adenovirus type 5 hexon and fiber vaccines. *Infect. Immun.* **25**, 771–773.
223. Finberg, R., Mescher, M. & Burakoff, S. J. (1978). The induction of virus-specific cytotoxic T lymphocytes with solubilized viral and membrane proteins. *J. Exp. Med.* **148**, 1620–1627.
224. Uchida, T., Kim, J., Yamaizumi, M., Miyake, Y. & Okada, Y. (1979). Reconstitution of lipid vesicles associated with HVJ (Sendai virus) spikes. *J. Cell Biol.* **80**, 10–20.
225. Huang, R. T. C., Wahn, K., Klenk, H.-D. & Rott, R. (1979). Association of the envelope glycoproteins of influenza virus with liposomes—a model study on viral envelope assembly. *Virology* **97**, 212–217.
226. Gerlier, D., Sakai, F. & Dore, J. F. (1978). Entrapment of the Gross virus associated cell surface antigen in liposomes. *C.R. Hebd. Séances Acad. Sci. Ser. D.* **286**, 439–442.
227. Dimitriadis, G. J. (1978). Entrapment of ribonucleic acids in liposomes. *FEBS Lett.* **86**, 289–293.
228. Wilson, T., Papahadjopoulos, D. & Taber, R. (1979). The introduction of poliovirus RNA into cells via lipid vesicles (liposomes). *Cell* **17**, 77–84.
229. Dimitriadis, G. J. (1978). Entrapment of plasmid DNA in liposomes. *Nucleic Acid Res.* **6**, 2697–2708.
230. Mukherjee, B., Orloff, S., Butler, J., Triche, I., Lalley, P. & Schulman, J. D. (1978). Entrapment of metaphase chromosomes into phospholipid vesicles (lipochromosomes). *Proc. Natl. Acad. Sci. U.S.A* **75**, 1361–1365.
231. Mannino, R. J., Allebach, E. S. & Strohl, W. A. (1979). Encapsulation of high molecular weight DNA in large unilamellar phospholipid vesicles. *FEBS Lett.* **101**, 229–232.
232. Lurquin, P. F. (1979). Entrapment of plasmid DNA by liposomes and their interaction with plant protoplasts. *Nucleic Acid Res.* **6**, 3773–3784.
233. Mountcastle, D. B., Biltonen, R. L. & Halsey, M. J. (1978). Effect of anesthetics and pressure on the thermotropic behaviour of multilamellar dipalmitoyl-phosphatidylcholine liposomes. *Proc. Natl. Acad. Sci. U.S.A.* **75**, 4906–4910.
234. Fraley, R. T., Fornari, C. S. & Kaplan, S. (1979). Entrapment of a bacterial plasmid in phospholipid vesicles: potential for gene transfer. *Proc. Natl. Acad. Sci. U.S.A.* **76**, 3348–3352.
235. Kamaya, H., Uede, I., Moore, P. S. & Eyring, H. (1979). Antagonism between high pressure and anesthetics in the thermal phase-transition of dipalmitoyl phosphatidylcholine bilayers. *Biochim. Biophys. Acta* **550**, 131–137.
236. MacDonald, A. G. (1978). A dilatometric investigation of the effects of general anesthetics, alcohols and hydrostatic pressure on the phase-transition in smectic mesophases of dipalmitoylphosphatidylcholine. *Biochim. Biophys. Acta* **507**, 26–37.

237. Rahamimoff, R., Meiri, H., Erulkar, S. D. & Barenholz, Y. (1978). Changes in transmitter release induced by ion-containing liposomes. *Proc. Natl. Acad. Sci. U.S.A.* **75**, 5214–5216.
238. Covarrubias, M. & Tapia, R. (1978). Calcium-dependent binding of brain glutamate decarboxylase to phospholipid vesicles. *J. Neurochem.* **31**, 1209–1214.
239. Gutman, Y., Lichtenberg, D., Cohen, J. & Boonyariroj, P. (1979). Increased catecholamine release from adrenal medulla by liposomes loaded with sodium or calcium ions. *Biochem. Pharmacol.* **28**, 1209–1211.
240. Theoharides, T. C. & Douglas, W. W. (1978). Secretion in mast cells induced by calcium entrapped within phospholipid vesicles. *Science* **201**, 1143–1145.
241. Dormer, R. L., Hallett, M. B. & Campbell, A. K. (1978). The incorporation of the calcium-activated photoprotein into isolated rat cells by liposome cell fusion. *Biochem. Soc. Trans.* **6**, 570–572.
242. Stendahl, O. & Tagesson, C. (1977). Interaction of liposomes with polymorphonuclear leukocytes. I. Studies on the mode of interaction. *Exp. Cell Res.* **108**, 167–174.
243. Dahlgren, C., Kihlström, E., Magnusson, K.-E., Stendahl, O. & Tagesson, C. (1977). Interaction of liposomes with polymorphonuclear leukocytes. II. Studies on the consequences of interaction. *Exp. Cell Res.* **108**, 175–184.
244. Kriss, J. P. & Mehdi, S. Q. (1979). Cell-mediated lysis of lipid vesicles containing eye muscle protein: implications regarding pathogenesis of Graves ophthalmology. *Proc. Natl. Acad. Sci. U.S.A.* **76**, 2003–2007.
245. Cassells, A. C. (1978). Uptake of charged lipid vesicles by isolated tomato protoplasts. *Nature (London)* **275**, 760.
246. Stillwell, W. & Tien, H. T. (1978). Oxygen evolution from broken thylakoids fused with liposomes. *Biochem. Biophys. Res. Commun.* **81**, 212–216.
247. Schwendener, R. A. & Weder, H.-G. (1978). The binding of chlorpromazine to bilayer liposomes. Evaluation of stoichiometric constants from equilibrium and steady state studies. *Biochem. Pharmacol.* **27**, 2721–2727.
248. Chattopadhyay, P. K., Lai, J.-S. & Wu, H. C. (1979). Incorporation of phosphatidylglycerol into murein lipoprotein in intact cells of *Salmonella typhimurium* by phospholipid vesicle function. *J. Bacteriol.* **137**, 309–312.
249. Curatolo, W., Yau, A. O., Small, D. M. & Sears, B. (1978). Lectin-induced agglutination of phospholipid/glycolipid vesicles. *Biochemistry* **17**, 5740–5744.
250. Lee, G., Consiglio, E., Habig, W., Dyer, S., Hardegree, C. & Kohn, L. D. (1978). Structure:function studies of receptors for thyrotropin and tetanus toxin: lipid modulation of effector binding to the glycoprotein receptor component. *Biochem. Biophys. Res. Commun.* **83**, 313–320.
251. Margolis, L. B. & Bergelson, L. D. (1979). Lipid-cell interactions. Induction of microvilli on the cell surface by liposomes. *Exp. Cell Res.* **119**, 145–150.
252. Zborowski, J., Roerdink, F. & Scherphof, G. (1977). Leakage of sucrose from phosphatidylcholine liposomes induced by interaction with serum albumin. *Biochim. Biophys. Acta* **497**, 183–191.
253. Sefton, B. M. & Gaffney, B. J. (1979). Complete exchange of viral cholesterol. *Biochemistry* **18**, 436–442.
254. Roerdink, F., Wisse, E., Morselt, H., van der Meulen, J. & Scherphof, G. (1977). Cellular distribution of intravenously injected protein-containing liposomes in the rat liver. In *Kupffer Cells and Other Liver Sinusoidal Cells* (Wisse, E. & Knook, D. L., eds.), pp. 263–272. Elsevier/North-Holland Biomedical Press, Amsterdam.

255. Scherphof, G., Roerdink, F., Waite, M. & Parks, J. (1978). Disintegration of phosphatidylcholine liposomes in plasma as a result of interaction with high density lipoproteins. *Biochim. Biophys. Acta* **542**, 296–307.
256. Scherphof, G., Morselt, H., Regts, J. & Wilschut, J. C. (1979). The involvement of the lipid phase transition in the plasma-induced dissolution of multilamellar phosphatidylcholine vesicles. *Biochim. Biophys. Acta* **556**, 196–207.
257. Hoekstra, D. & Scherphof, G. (1979). Effect of fetal calf and serum protein fractions on the uptake of liposomal phosphatidylcholine by rat hepatocytes in primary monolayer culture. *Biochim. Biophys. Acta* **551**, 109–121.
258. Scherphof, G., Roerdink, F., Hoekstra, D., Zborowski, J. & Wisse, E. (1979). Stability of liposomes in presence of blood constituents: consequences for uptake of liposomal lipid and entrapped compounds by rat liver cells. In *Liposomes in Biological Systems* (Gregoriadis, G. & Allison, A. C., eds.), pp. 179–209. John Wiley & Sons, Chichester.
259. Hoekstra, D., van Renswoude, J., Tomasini, R. & Scherphof, G. (1980). Interaction of phospholipid vesicles with rat hepatocytes. Further characterization of vesicle-cell surface interaction; use of serum as a physiological modulator. *Membr. Biochem.*, in press.
260. Tall, A. R. & Lange, Y. (1978). Incorporation of cholesterol into high density lipoprotein recombinants. *Biochem. Biophys. Res. Commun.* **80**, 206–212.
261. Tall, A. R., Hogan, V., Askinazi, L. & Small, D. L. (1978). Interaction of plasma high density lipoproteins with dimyristoyllecithin multilamellar liposomes. *Biochemistry* **17**, 322–326.
262. Jonas, A. & Maine, G. T. (1979). Kinetics and mechanism of phosphatidylcholine and cholesterol exchange between single bilayer vesicles and bovine serum high-density lipoprotein. *Biochemistry* **18**, 1722–1728.
263. Ginsberg, L. (1978). Does Ca^{2+} cause fusion or lysis of unilamellar lipid vesicles? *Nature (London)* **275**, 758–760.
264. Heath, T. D., Robertson, D., Birbeck, M. S. C. & Davies, A. J. S. (1979). Liposomal fusion as a means of introducing surface antigens into living cell membranes. *GANN Monogr. Cancer Res.* **23**, 255–264.
265. Mauk, M. R. & Gamble, R. C. (1979). Stability of lipid vesicles in tissues of the mouse: A γ-ray perturbed angular correlation study. *Proc. Natl. Acad. Sci. U.S.A.* **76**, 765–769.
266. Hnatowic, D., Clancy, B., Kulprath, S. & O'Connell, T. (1979). New liposome type with improved distribution for imaging. *J. Nucl. Med.* **20**, 680–681.
267. Gregoriadis, G. & Neerunjun, D. E. (1974). Control of the rate of hepatic uptake and catabolism of liposome-entrapped proteins into rats. Possible therapeutic application. *Eur. J. Biochem.* **47**, 179–185.
268. Richardson, V. J., Jeyasingh, K., Jewkes, R. F., Ryman, B. E. & Tattersall, M. H. N. (1977). Properties of [^{99m}Tc]technetium-labelled liposomes in normal and tumour-bearing rats. *Biochem. Soc. Trans.* **5**, 290–291.
269. Schieren, H., Weissmann, G., Seligman, M. & Coleman, P. (1978). Interactions of immunoglobulins with liposomes: an ESR and diffusion study demonstrating protection by hydrocortisone. *Biochem. Biophys. Res. Commun.* **82**, 1160–1167.
270. Huang, L. & Kennel, S. J. (1979). Binding of immunoglobulin G to phospholipid vesicles by sonication. *Biochemistry* **18**, 1702–1707.
271. Torchilin, V. P., Goldmacher, V. S. & Smirnov, V. N. (1978). Comparative studies on covalent and non-covalent immobilization of protein molecules on the surface of liposomes. *Biochem. Biophys. Res. Commun.* **85**, 983–990.

272. Weinstein, J. N., Blumenthal, R., Sharrow, S. O. & Henkart, P. A. (1978). Antibody-mediated targeting of liposomes. Binding to lymphocytes does not ensure incorporation of vesicle contents into the cells. *Biochim. Biophys. Acta* **509**, 272–288.

273. Leserman, L. D., Weinstein, J. N., Blumenthal, R., Sharrow, S. O. & Terry, W. D. (1979). Binding of antigen-bearing fluorescent liposomes to the murine myeloma tumor MOPC 315. *J. Immunol.* **122**, 535–591.

274. Trouet, A., Deprez-de Campeneere, D., Baurain, R., Huybrechts, M. & Zenebergh, A. (1979). Desoxyribonucleic acid as carrier of antitumor drugs. In *Drug Carriers in Biology and Medicine* (Gregoriadis, G., ed.), pp. 87–105. Academic Press, London and New York.

275. Shier, W. T. (1979). Lectins as drug carriers. In *Drug Carriers in Biology and Medicine* (Gregoriadis, G., ed.), pp. 47–70. Academic Press, London and New York.

276. Dean, R. T. (1979). The carrier potential of some glycoproteins, In *Drug Carriers in Biology and Medicine* (Gregoriadis, G., ed.), pp. 71–86. Academic Press, London and New York.

277. Molteni, L. (1979). Dextrans as drug carriers. In *Drug Carriers in Biology and Medicine* (Gregoriadis, G., ed.), pp. 107–125. Academic Press, London and New York.

278. Yatvin, M. B., Weinstein, J. N., Dennis, W. H. & Blumenthal, R. (1978). Design of liposomes for enhanced local release of drugs by hyperthermia. *Science* **202**, 1290–1293.

279. Weinstein, J. N., Magin, R. L., Yatvin, M. B. & Zaharko, D. S. (1979). Liposome and local hyperthermia: selective delivery of methotrexate to heated tumors. *Science* **204**, 188–191.

280. Gilliland, D. G., Collier, R. J., Moehring, J. M. & Moehring, T. J. (1978). Chimeric toxins: Toxic, disulfide-linked conjugate of concanavalin A with fragment A from diphtheria toxin. *Proc. Natl. Acad. Sci. U.S.A.* **75**, 5319–5323.

281. Yamaguchi, T., Kato, R., Beppu, M., Terao, T., Inoue, Y., Ikawa, Y. & Osawa, T. (1979). Preparation of concanavalin A-ricin A-chain conjugate and its biologic activity against various cultured cells. *J. Natl. Cancer Inst.* **62**, 1387–1395.

282. Uchida, T., Yamaizumi, M., Mekadd, E., Okada, Y., Ysuda, M., Kurokawa, T. & Sugino, Y. (1978). Reconstitution of hybrid toxin from fragment of diphtheria toxin and subunits of *Wisteria floribunda* lectin. *J. Biol. Chem.* **253**, 6307–6310.

283. Oeltman, T. N. & Heath, E. C. (1979). A hybrid protein containing the toxic subunit of human chorionic gonadotrophin. *J. Biol. Chem.* **254**, 1022–1027.

284. Thorpe, P. E., Ross, W. C. J., Cumber, A. J., Hinson, C. A., Edwards, D. C. & Davies, A. J. S. (1978). Toxicity of diphtheria toxin for lymphoblastoid cells is increased by conjugation to antilymphocytic globulin. *Nature (London)* **271**, 752–755.

285. Dimitriadis, G. J. & Butters, T. D. (1979). Liposome-mediated ricin toxicity in ricin-resistant cells. *FEBS Lett.* **98**, 33–36.

286. Gardas, A. & McPherson, I. (1979). Microinjection of ricin entrapped in liposomes into a ricin-resistant mutant of baby hamster kidney cells. *Biochim. Biophys. Acta* **584**, 538–541.

287. Ihler, G. M. (1979). Potential use of erythrocytes as carriers for enzymes and drugs. In *Drug Carriers in Biology and Medicine* (Gregoriadis, G., ed.), pp. 129–153. Academic Press, London and New York.

288. Zimmerman, U., Pilwat, G. & Esser, B. (1978). The effect of encapsulation in red blood cells on the distribution of methotrexate in mice. *J. Clin. Chem. Clin. Biochem.* **16**, 135–144.
289. Segal, A. W. (1979). Neutrophilic polymorphonuclear leucocytes. In *Drug Carriers in Biology and Medicine* (Gregoriadis, G., ed.), pp. 155–165. Academic Press, London and New York.
290. Harris, G. (1979). Lymphoid cells and transport of macromolecules. In *Drug Carriers in Biology and Medicine* (Gregoriadis, G., ed.), pp. 167–190. Academic Press, London and New York.
291. Chang, T. M. S. (1979). Artificial cells as drug carriers in biology and medicine. In *Drug Carriers in Biology and Medicine* (Gregoriadis, G., ed.), pp. 271–285. Academic Press, London and New York.
292. Wise, D. L., Fellman, T. D., Sanderson, J. E. & Wentworth, R. L. (1979). Lactic/glycolic acid polymers. In *Drug Carriers in Biology and Medicine* (Gregoriadis, G., ed.), pp. 237–270. Academic Press, London and New York.
293. Widder, K., Flouret, G. & Senyei, A. (1979). Magnetic microspheres: synthesis of a novel parenteral drug carrier. *J. Pharm. Sci.* **68**, 79–82.
294. Mosbach, K. & Schröder, U. (1979). Preparation and application of magnetic polymers for targetting of drugs. *FEBS Lett.* **102**, 112–117.
295. Marty, J. J., Oppenheim, R. C. & Speiser, R. (1978). Nanoparticles—a new colloidal drug delivery system. *Pharm. Acta Helv.* **53**, 17–23.

Steroid Hormone Production by Mammalian Adrenocortical Dispersed Cells

J. F. TAIT, S. A. S. TAIT and J. B. G. BELL

Biophysical Endocrinology Unit, Department of Physics as Applied to Medicine, Middlesex Hospital Medical School, London W1P 6DB, England

Students who do not wish to study the subject in depth are advised to read Sections I and IV and the italicized parts of Section III.

Abbreviations:

ZG cells, zona glomerulosa cells
ZF cells, zona fasciculata cells
ZR cells, zona reticularis cells
ACTH, adrenocorticotropin
FSH, follicle stimulating hormone
LH, luteinizing hormone
HCG, human chorionic gonadotropin

cyclic AMP, adenosine 3′,5′-cyclic monophosphate
cyclic GMP, guanosine 3′,5′-cyclic monophosphate
MSH, melanocyte stimulating hormone
MIX, 3-isobutyl-1-methylxanthine
NPS derivative, o-nitrophenyl sulphenyl derivative

I. Introduction to Adrenocortical Hormones

This introduction will describe briefly the state of knowledge in the field up to the time when dispersed adrenocortical cells came to be used extensively. The earliest applications were for the zona fasciculata (ZF), then zona glomerulosa (ZG) and quite recently for zona reticularis (ZR) cells.

A. NATURE OF SECRETION AND BIOSYNTHESIS

In the 1950s the unitarian hypothesis of the nature of the secretion of the mammalian adrenal cortex finally collapsed.[1] It was then accepted that the composition of the total steroid production of the gland was complex. Different steroids are responsible for the major effects on certain aspects of electrolyte (mineralocorticoid), carbohydrate (glucocorticoid) and anabolic (androgenic) metabolism. Aldosterone, in all mammalian species studied, is responsible for most if not all the mineralocorticoid activity (Table 1). Cortisol, in man and most other mammalian species, contributes significantly to mineralocorticoid activity and is responsible for all the glucocorticoid activity. Corticosterone is produced by all mammalian species, and usually in appreciable quantities, but its function, when cortisol is secreted as in man, is not clear. It does not serve as a biosynthetic precursor for cortisol, as might seem likely from its chemical

TABLE 1

Production of steroids by the zonae glomerulosa, fasciculata and reticularis in different species. When cortisol is not produced, corticosterone is made in greater amounts by the zona fasciculata-reticularis than indicated

Steroid	Zona glomerulosa	Zona fasciculata-reticularis	Species
Cortisol	−	+ + + +	Major glucocorticoids in most mammals including man; excluding rat, mouse, rabbit and ferret
11-Deoxycortisol	−	+	
18-Hydroxy deoxycorticosterone	+	+ + +	All species examined
Corticosterone	+ +	+ + +	Major glucocorticoid in rat, mouse, rabbit and ferret
Deoxycorticosterone	+	+ +	All species examined
18-Hydroxy corticosterone	+ +	+	All species examined
Aldosterone	+	−	Major mineralocorticoid in all mammals examined

structure[2] (Fig. 1). Steroids such as 18-hydroxycorticosterone and 18-hydroxydeoxycorticosterone[3, 4] are also secreted in appreciable amounts by most species, but their physiological function and even their role in pathological conditions has not been defined (Table 1). Other steroids such as deoxycorticosterone (with mainly salt retaining activity), progesterone and 17α-hydroxyprogesterone (with mainly salt losing activity) may have biological effects in certain pathological conditions resulting from enzymic defects,[5] but their physiological roles are probably minor. Androgenic steroids secreted by the adrenal cortex are androstenedione, 11β-hydroxy-androstenedione, dehydroepiandrosterone and dehydroepiandrosterone sulphate. All these are intrinsically only weak androgens but they can be converted peripherally to more potent androgens, such as testosterone and dihydrotestosterone, which could be important physiologically in the female.[6] Androstenedione can also be converted to oestrogens such as oestrone but the physiological significance of such conversions is still unknown. Because of a prior use of the term prohormone by Emmens in a rather different sense,[6] it has been customary in the steroid field to refer to such secreted peripheral precursors as prehormones.

The accepted major biosynthetic pathways for the steroid hormones within the adrenal cortex are shown in Fig. 1. When putting forward such a scheme, it must be realized that there are several alternative pathways. It is very difficult, even in narrowly defined situations, to delineate quantitatively the relative importance of multiple pathways leading to a common product. A significant alternative pathway from cholesterol to cortisol and other Δ^4-3-oxo steroids has been demonstrated which proceeds not through progesterone but through intermediates such as pregnenolone,[7] which retain the 3β-hydroxy-5-ene structure until the later stages of biosynthesis. Furthermore it has been suggested[8] that androstenedione in species such as the rat, which lack 17α-hydroxylase, could be formed from progesterone via testosterone acetate. It seems likely also that there is an important pathway to aldosterone not passing through corticosterone but using 18-hydroxydeoxycorticosterone as an intermediate.[9] Even when corticosterone is the precursor, it seems possible that the pathway proceeds not through 18-hydroxycorticosterone but through a hydroperoxide.[10] Such an intermediate in the conversion of cholesterol to pregnenolone has been suggested by Lieberman and coworkers.[11]

It should be emphasized that, for the sake of simplicity, the biosynthetic scheme illustrated (Fig. 1) shows common precursors for all products in the three types of cell. This could give rise to the false conclusion that, in the case of tropic effects of humoral substances or ions or in the presence of enzymic defects, accumulation of precursors in one zone would increase the secretion of all products. However, if these are synthesized by different regions of the adrenal cortex, this conclusion would depend on the microcirculation of these

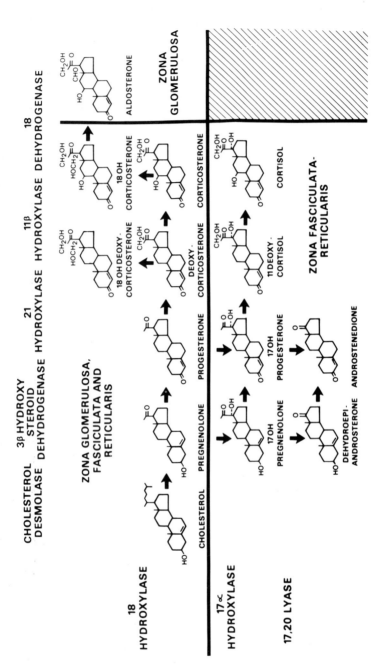

Fig. 1. Major biosynthetic pathways for adrenal steroids in zonae glomerulosa, fasciculata and reticularis. Common pathways shown for the three zones (top left) are shared conceptually only and do not actually involve the same pools of steroids.

glands and the systemic metabolic clearance rates of the precursors and products. Only when a complete knowledge of the site of production of all significant steroids is achieved will it be possible to dissect such schemes on a zonal basis to give more reliable predictions of the secretion rates of steroids in various physiological and pathological situations.

B. SITE OF PRODUCTION

The histology of the mammalian adrenal cortex shows three major groups of steroidogenically active cell characterized by their organization *in situ*;[12] the zona glomerulosa (outer region: ball pattern), the zona fasciculata (central region: columnar pattern) and the zona reticularis (inner region: network pattern). There are other types of cell between these major zones, e.g. the zona intermedia comprising cells between the zona glomerulosa and fasciculata, and these may have interesting properties.[12] The composition of the zona reticularis in the rat gland and in certain other species is complex and consists of both the most numerous cells with low lipid (compact) and of small diameter (about 8 μm) but also some high lipid (clear) cells of similar diameter (Fig. 2). The

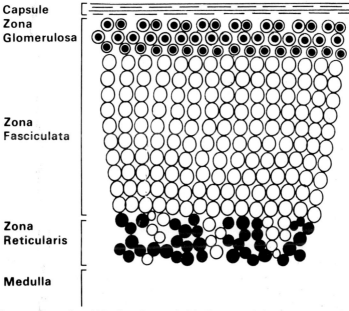

Fig. 2. Diagram of rat adrenal histology (not to scale). The open circles represent small cells in the zona reticularis region with high lipid content and spherical cristae characteristic of ZF cells. The fully shaded ZR cells (major component) are also small but have low lipid content and unique mixed tubular-spherical cristae.

cells of the zona fasciculata have a high lipid content (clear cells) and are larger (about 18 μm mean diameter).[5, 13] The electron microscopic appearance of the mitochondria, with spherical cristae, of these zona fasciculata cells is similar to that of the smaller clear cells in the zona reticularis. Fingers of such clear cells run from the zona fasciculata through the zona reticularis to the medulla (Fig. 2). The mitochondria of the compact cells show a characteristic mixture of spherical and tubular cristae, whereas those of zona glomerulosa cells show only tubular cristae. Therefore the three major types of cells can be distinguished by electron microscopy either *in situ* or after dispersion.[13]

If the capsule with adhering tissue is stripped from the rat adrenal gland by a technique pioneered by Giroud and coworkers,[14] the active steroidogenic cells of the resultant capsular tissue can consist of up to 95% zona glomerulosa with only 5% zona fasciculata cells. The cortical tissue of the remaining decapsulated gland consists of a mixture of zona fasciculata and reticularis cells (usually in a 1 : 1 ratio). Nevertheless these two impure preparations have proved useful in indicating the site of production of steroids in the rat adrenal cortex. In glands from other species such as the ox, similar separation of the zona glomerulosa from the zona fasciculata-reticularis can be achieved by preparing slices using a Stadie–Rigg microtome.[15] Although this procedure is less successful in the case of the convoluted human adrenal, some indication of zonal function can still be obtained.[16]

Application of these techniques has demonstrated that aldosterone is produced only by the zona glomerulosa and cortisol, in those species producing it, only by the zona fasciculata-reticularis. Corticosterone is produced by both the capsular and decapsulated rat adrenals and also by the equivalent preparations from other species whether they secrete cortisol or not and hence is synthesized by both the zonae glomerulosa and fasciculata-reticularis. 18-Hydroxycorticosterone and 18-hydroxydeoxycorticosterone may be produced preferentially by one or other of these regions (usually 18-hydroxycorticosterone by the zona glomerulosa and 18-hydroxydeoxycorticosterone by the zona fasciculata-reticularis) but they are definitely produced by both zonae glomerulosa and fasciculata-reticularis.[17, 18]

The problem of the differential production of steroids by the zonae fasciculata and reticularis is more difficult to resolve using adrenal tissue only. However, employing dissection of frozen adrenal slices Griffiths & Glick[19] obtained the first direct results. Vines[20] had previously concluded that androgens were produced preferentially by the zona reticularis but this was based mainly on indirect correlations between histology and biological activity in human pathological conditions. Jones & Griffiths[21] found that core samples from the guinea-pig adrenal cortex, consisting mainly of zona reticularis cells, synthesized dehydroepiandrosterone sulphate from dehydroepiandrosterone, whereas similar samples containing mainly zona fasciculata tissue were much less active in this respect. This study has been the basis for the assumption by several groups that secretion of dehydroepiandrosterone sulphate in man is an indication of zona reticularis activity. However, endogenous production of dehydroepiandrosterone and its sulphate was not measured in the guinea-pig studies. It seems likely that only the higher primates secrete dehydroepiandrosterone and its sulphate in appreciable amounts.[22] Ward & Grant[23] using adrenocortical tissue slices found that, in man, both the zonae fasciculata and reticularis produced androstenedione. However, as the authors admitted, these preparations contained mixtures of cells so that no firm conclusions could be drawn.

C. NATURE OF CONTROL

The nature of the control of corticosteroid secretion by cells of the zona fasciculata appears to be the most straightforward of all the mechanisms found in the adrenal cortex. ACTH seems to be the only important humoral factor controlling zona fasciculata secretion. The failure to isolate a releasing factor for ACTH (CRF) has obscured the mechanisms of hypothalamic control, although the overall situation seems to be relatively simple. Nevertheless, it has to be realized that the usual preparations of adrenal tissue used in most of these studies were a mixture of two types of steroidogenic cell, those of the zona fasciculata and of the zona reticularis. Isolation and study of pure zona fasciculata and reticularis tissue might have transformed this apparently simple picture. The core technique employed by Griffiths & Glick[19] and by the Australian group of Coghlan and co-workers[24] could have provided material for more revealing studies on adrenal tissue, but it seems that technical problems and the relatively poor viability of sheep preparations *in vitro* has so far prevented the application of this approach to provide definitive answers to the major questions.

The nature of the control of androgen secretion by the adrenal is more complex. This is very difficult to analyse from measurements of plasma steroid concentrations in man because of the different sites of production of hormones and precursors and their interconversion peripherally. Dehydroepiandrosterone is converted unidirectionally to androstenedione which is itself transformed with interconversion into testosterone peripherally. Dehydroepiandrosterone sulphate is also interconverted to dehydroepiandrosterone. Although techniques are available to measure the interconversion and production rates of these steroids,[25] the calculation of secretion rates often involves considerable errors and assumptions and cannot be carried out continuously during stimulation experiments. In these circumstances it is essential to understand the nature of the control of adrenal androgens from *in vitro* studies.

Physiological studies indicate that these androgens are to some extent under ACTH control. However, the increase in adrenal androgens during adrenarche occurs somewhat earlier than most pubertal changes without any corresponding increase in cortisol.[26] Moreover, Parker & Odell[27] have recently postulated, from *in vivo* studies on castrated dogs, the presence of a cortical androgen-stimulating substance in bovine pituitary extract. Their studies indicated a higher stimulation of the adrenal androgens, dehydro-epiandrosterone and androstenedione, compared with cortisol for the pituitary extract than for either [1-39]ACTH or [1-24]ACTH. Prolactin, luteinizing hormone-releasing hormone, growth hormone, thyroid stimulating hormone and arginine vasopressin did not increase adrenal androgens under these conditions. It has been postulated, for this and other reasons, that a pituitary hormone other than ACTH may control the production of adrenal androgens although the presence of ACTH may also be necessary.[26] However, as Anderson has suggested (personal communication), this could be the result of a different mode of stimulation by ACTH, e.g. a different pulse frequency of ACTH stimulation. Other studies including those carried out *in vitro* on adrenal tissue indicate that free dehydroepiandrosterone and androstenedione may be produced by both zonae fasciculata and reticularis but dehydroepiandrosterone sulphate only by zona reticularis.[21] Steroidogenesis by the two types of cell could be under different modes of control, and the role of pituitary hormones other than ACTH, e.g. gonadotropins and prolactin, is still largely unresolved. It is therefore evident that *in vitro* studies on pure cells are essential to establish the site and nature of androgen production in various species.

The nature of the control of aldosterone secretion has been one of the most intensive areas of endocrine research during the last twenty-five years. It has emerged that the renin–angiotensin system is a major factor in the control of aldosterone in all the mammalian species studied with the direct influence of K^+ concentration important in certain situations. ACTH can also affect aldosterone secretion but usually the effects are transient.[28] Serotonin can affect steroid production by rat adrenal tissue *in vitro*[29] and, although meaningful effects *in vivo* have not yet been established, recent preliminary results in humans and rats by Edwards and coworkers[30, 31] seem convincing. Other *in vivo* factors have been proposed but these have not been definitively established by the isolation of the appropriate humoral substances although recently there have been reports of partially characterized glycoproteins from human urine which seem to be active and specific stimulators of aldosterone.[32, 33] Palmore & Mulrow[34] have reported that, in the rat, a pituitary factor is essential for aldosterone control. This has not been identified although Page *et al.*[35] and more recently Vinson *et al.*[36] reported that α-MSH was weakly active. Reported discrepancies in the effects of angiotensin in the rat *in vivo* have been shown to have arisen from the use of anaesthetics in earlier studies.[37] Later it was shown that the rat differed from other species in that anaesthetics stimulated the renin–angiotensin system so that no apparent effect was seen on subsequent administration of exogenous angiotensin. Studies on rat adrenal tissue also showed little activity of angiotensin II in stimulating aldosterone production.

The outstanding problem in the acceptance of the renin–angiotensin system as the major component in the effect of salt depletion on aldosterone secretion is that the angiotensin concentration in animals on a normal sodium intake has to be greater than in salt depletion in order to achieve comparable aldosterone levels. It has therefore been necessary to postulate that the adrenal in salt depletion is sensitized to the effects of angiotensin,[38] a view now generally accepted.[39] However unless a specific mechanism, preferably on a molecular basis, is postulated, the position is not more advanced than is the suggestion that an unknown mechanism other than the renin–angiotensin system is operative. If angiotensin itself increased the sensitivity of the gland by a specific mechanism, such as effects on the late pathway for aldosterone synthesis, then the whole nature of the major control could be explained. However, the Glasgow group of Lever and co-workers[40] have shown that, while the infusion of angiotensin II increases the sensitivity of the adrenal, the response still falls short of that which can be achieved by salt depletion (Section III.B.2).

D. MECHANISM OF CONTROL

The seemingly simple nature of the control of steroidogenesis by zona fasciculata cells with ACTH as the only important tropic humoral factor, and with no direct external ionic effects under physiological conditions, is matched by there being initially only one viable theory to explain the mechanism of ACTH action. This is the Sutherland second messenger hypothesis involving cyclic AMP[41]. The first version of this theory as applied to the adrenal cortex by Haynes & Berthet[42] proposed that the nucleotide increased steroidogenesis by raising phosphorylase activity and NADPH availability but subsequent work showed this to be unlikely[43] at least as a major mechanism. Thus Harding & Nelson[44] found that in hypophysectomized rats the concentration of NADPH remained constant, regardless of stimulation with ACTH.

The model of Garren and co-workers[45] proved to be more in accordance with experimental evidence, both *in vivo* and *in vitro* with adrenal tissue. This involved binding of ACTH to the plasma membrane with stimulation of adenylate cyclase and cyclic AMP production. Cyclic AMP-activated protein kinases increase the phosphorylation of several proteins perhaps including ribosomal protein which may regulate protein synthesis at the translational level. Evidence in support of this has been provided in the studies of Rae *et al.*[46]

who, by using mutant clones, showed that adenylate cyclase and cyclic AMP-dependent kinase were obligatory components of steroid stimulation by ACTH.

Using quartered rat adrenal glands, Grahame-Smith et al.[41] showed good correlation between steroidogenesis and cyclic AMP production in response to stimulation by ACTH and ACTH analogues. Studies by Schulster et al.[47] using superfused rat decapsulated adrenal glands and stimulation of steroidogenesis by cyclic AMP and ACTH showed that the data were consistent with an intermediary role for cyclic AMP. However, a number of studies with endocrine tissue, including adrenal, demonstrated a dichotomy between cyclic AMP and steroid production after stimulation with high doses of tropic hormones. This led to the formulation of the "spare receptor" theory which may be an important factor in increasing the sensitivity of the target cells to stimulation by tropic hormones.[48]

Stone & Hechter[49] first demonstrated that ACTH acted in the biosynthetic pathway between cholesterol and pregnenolone and Garren and co-workers[45] and Karaboyas & Koritz[50] showed that cyclic AMP stimulated the same step. Koritz & Hall[51] proposed that the rate limiting step in this stimulation of steroidogenesis was the rate at which the synthesized pregnenolone was removed from the mitochondria and that cyclic AMP, acting on the permeability of the mitochondrial membrane, facilitated this removal. This view is no longer favoured because mitochondrial pregnenolone levels are increased rather than decreased after ACTH stimulation.[52] There now appears to be reasonable agreement that ACTH stimulation results from increased amounts of cholesterol forming a substrate complex with cytochrome P-450 and not to an increase in the intrinsic activity of the enzyme systems responsible for side chain cleavage. Theoretically, this could be due to increased cholesterol esterase activity, transport of free cholesterol from the lipid stores to the mitochondria and/or binding of the cholesterol to the P-450 system.

The process stimulated by ACTH is sensitive to cycloheximide,[47] which might inhibit the formation of a labile protein involved either in transferring cholesterol into the mitochondria or in the interaction of cholesterol with cytochrome P-450. The ACTH-induced activation of cholesterol esterase, which is cyclic AMP-dependent and is mediated by a protein kinase-dependent phosphorylation, appears to be cycloheximide insensitive.[52] Pederson & Brownie[53] reported that ACTH does not activate the esterase. The observed effect of stress on the esterase is probably mediated by a pituitary factor related to γ-MSH. The overall evidence favours the view that the major step activated by ACTH is the increased association of the mitochondrial cholesterol with the cytochrome P-450[51] and does not involve transport of cholesterol into the mitochondria.[52]

Studies in vivo and in vitro on adrenal tissue have established that Ca^{2+} is implicated at several steps in steroidogenic stimulation. Although Ca^{2+} is not required for attachment of ACTH to adrenal receptors,[54] optimal concentrations are necessary for ACTH activation of adenylate cyclase.[55] Protein synthesis which is involved in steroidogenic stimulation by ACTH requires the presence of Ca^{2+} and the transport of cholesterol and its attachment to the cytochrome P-450 may also be facilitated by Ca^{2+}.[52]

There have been reports[56] that ACTH may also act on later stages of biosynthesis, such as on 11β-hydroxylation, but these results may also be due to chronic, non-specific effects in which many enzymes are increased or to "artefactual" phenomena involving alterations in peripheral pathways, e.g. decreased conversion of 11β-hydroxy to 11-oxo steroids (see Section III.C.1).

Therefore before work on dispersed cells was started, the mechanism of action of ACTH seemed to be reasonably established, although the rapid inhibitory effect of substances such as cycloheximide on the stimulation of steroidogenesis by ACTH and hence the concomitant obligatory role of protein synthesis was still unexplained. Also, all these earlier *in vitro* studies were carried out with a mixture of zona fasciculata and reticularis cells and it remained to be seen whether the use of pure zona fasciculata cells would affect the interpretation of the results.

With the lack of knowledge of the nature of control of the zona reticularis cells (indeed one influential review described the zone as vegetative and senescent[57]), no experimental work had been carried out on the mechanism of action of their stimulators. It was suspected that they were under the influence of ACTH and that this might act according to the Sutherland–Garren model, but no relevant definitive evidence had been reported. This had to await the availability of purified zona reticularis cells.

The use of capsular preparations of rat adrenals and corresponding tissue preparations from other species, containing a high proportion of zona glomerulosa cells, has enabled fruitful work on the mechanism of action of humoral and ionic stimulators to be accomplished. These preparations contained at least 5% contamination with zona fasciculata cells and this can lead to confusing interpretations particularly when the contaminating cells are stimulated, as will be described later. However, this was a minor consideration in most earlier work.[58]

In contrast to the results of studies on the stimulation of steroid biosynthetic pathways in the zona fasciculata, a feature of the results of corresponding studies on the zona glomerulosa was that, in addition to stimulation of the cholesterol to pregnenolone step, there were specific effects on late pathways such as corticosterone to aldosterone and probably also 18-hydroxycortico-sterone to aldosterone. Both *in vitro* and *in vivo* studies have demonstrated effects of all known aldosterone stimulators on the early pathway. Effects of salt depletion on the late pathway have been demonstrated in the rat within

48 h when the zona glomerulosa width is not markedly increased.[59] A similar effect, probably acting specifically on the zona glomerulosa, has been demonstrated in the sheep adrenal transplant.[60] Even more convincing was the work of Marusic and co-workers[61] who showed an effect of renin in the dog within 3 h, and recently Brownie and co-workers[62] have reported an effect on a late pathway cytochrome P-450 system not only with long term salt depletion but also 30 min after the administration of angiotensin II.[63] K^+ administration can also activate the late pathway.[28,64] It is necessary in many of these long-term experiments to control carefully the sodium balance otherwise the externally administered substance may be acting by causing natriuresis through, for example, direct effects on the kidney leading to other mechanisms for stimulating aldosterone. Also it has been reported that renin in the rat and K^+ administration in the dog do not affect the late pathway (Marusic, personal communication), although Komor & Muller[28] have shown a small but significant effect of angiotensin II infusion on this pathway in the rat. Nevertheless, it is possible *in vivo* to demonstrate in most species an effect on the early and late pathway by salt depletion and by other factors such as angiotensin II and K^+ concentration postulated to be the appropriate physiological messengers. With *in vitro* tissue studies, although effects on the early pathway could be readily demonstrated, the late pathway could not be shown to be stimulated by the addition of substances such as angiotensin II and K^+. This is distinct from the demonstration *in vitro* of an effect from *in vivo* stimulation which can be readily shown.[59,64] It was the early difficulties of demonstrating effects on the late pathway which led to such biosynthetic evidence being quoted in support of there being an unknown physiological messenger substance which could have an effect on the late pathway.[60] However, there were complicating factors such as kinetic considerations (salt depletion is usually achieved over a longer time period than is required for the action of specific humoral or ionic influences, particularly *in vitro*) and the question arises as to whether *in vitro* conditions are appropriate in other respects for the demonstration of effects of postulated physiological messengers such as angiotensin II or K^+ concentration.

Some of the earliest studies on the role of cyclic AMP in the stimulation of aldosterone production by adrenal tissue were carried out by Kaplan[65] using ox adrenal slices. He found that added cyclic AMP did not increase further the maximal stimulation by ACTH where there was additional stimulation after maximal effects of angiotensin II. However, little work was then done with measurements of cyclic AMP after stimulation of adrenal tissue.

The mechanism of stimulation of steroidogenesis by small changes in K^+ concentration was of great interest, not only from the point of view of adrenal endocrinology but also because of the implications for understanding corresponding biophysical phenomena generally. Relevant studies on adrenal tissue showed that Ca^{2+} was necessary for stimulation of aldosterone including that by K^+ and angiotensin II,[58] and moreover ammonium, rubidium and caesium ions could mimic the action of K^+.[58] Ouabain was found either to inhibit or to stimulate aldosterone production, depending on the concentration of the glycoside used,[66] as has also been found for its effects on the production of substances by other tissues. Wellen & Benraad[67] found that 10^{-7} M ouabain inhibited aldosterone production by calf adrenals and simultaneously blocked active cation transport across the cell membranes resulting in loss of intracellular K^+. Muller[58] found that a dose of ouabain, which stimulated basal aldosterone production by 40%, completely inhibited the increase in steroid formation due to K^+ and serotonin and partially blocked the effect of ACTH. Coghlan and co-workers[68] investigated the effect of ouabain on the steroid and K^+ outputs of sheep adrenal transplants. They were unable to demonstrate a correlation between the effects on adrenal venous plasma electrolytes and steroid secretion. Plasma K^+ concentrations decreased at low doses and increased when the glycoside was raised to 10^{-5} M. By contrast aldosterone secretion was not consistently altered during the ouabain infusion but was markedly depressed after cessation of the infusion, whereas corticosterone and cortisol secretion rates increased. The authors conclude that these effects are not mediated by changes in intracellular K^+. The results are important as regards the effects of the glycoside in a physiological environment but have the problem that the K^+ and steroid outputs originate from all cells of the gland. A transplanted zona glomerulosa is required for this type of experiment.

Prostaglandins may also be involved in the control of aldosterone secretion. Thus indomethacin and other inhibitors of prostaglandin synthesis result in a decrease in plasma aldosterone levels in patients with Bartter's syndrome.[69] The effects are the consequence of prostaglandins acting on renin release rather than directly on the adrenal gland. Nevertheless Campbell et al.[70] and Spat & Jozan,[71] from studies in rats, have suggested that prostaglandins also have an effect directly on zona glomerulosa tissue (See Section III.B.1).

II. Aims and Methodology with Dispersed Cells

A. AIMS AND HISTORY

The aims of the use of dispersed cells are:

(i) To provide sources of cells for studies in long term culture, although it must be recognized that cultured cells do not always produce the same steroids as are secreted *in vivo* and their responses to known hormones are often abnormal.[72] Cultured preparations of ZG cells seem to be particularly deficient in reproducing physiological conditions.[73] ZR and ZF cells lose their differentiation after 24 h of culture.[74]

(ii) To improve the supply of nutrients and oxygen compared with tissue slices.

(iii) To improve access of tropic humoral factors and therefore increase the threshold sensitivity of response.

(iv) Although this was not intentional, to lower basal production rates compared with the corresponding tissue which again increases threshold sensitivity and also maximum stimulation ratios (stimulated/basal outputs).

(v) To enable purified homogeneous cells to be prepared by physical or other methods.

The technical disadvantage of the use of dispersed cells is the increased difficulty of carrying out morphological studies such as electron microscopy and of performing superfusion experiments which have proved to be useful with endocrine tissues. However, morphological techniques can be devised to overcome the difficulties.[75] Also, although a method of superfusion in columns, which was developed by Lowry & McMartin,[76] has the disadvantage that the cells are in contact with a solid support (Biogel or Sephadex) which potentially can adsorb tropic substances, it has, in practice, provided a valid approach for many situations, e.g. stimulation of superfused ZG cells by K^+ and serotonin.[77,78] The technique of Schulster,[79] which superfuses the cells in suspension in an elutriation apparatus without contact with solid support, offers an even better solution if certain technical problems can be overcome. A more fundamental theoretical objection to the use of dispersed cells is the removal of communication between cells which may be important *in situ*,[80] e.g. for the intercellular transport of cyclic AMP. Gap junctions between the cells are found in the adrenal cortex, particularly in the zona reticularis.[80] The transport of messengers such as cyclic AMP through such junctions is one possible explanation of the apparently increased sensitivity to tropic factors of cytochemical assays using tissue compared with those employing dispersed cells,[81] although carefully controlled experiments using the same source of cells and endpoint are required to examine this aspect. Results with tissue culture of adrenocortical cells have shown that responsiveness to ACTH is dependent on the density of the cell culture, possibly due to the necessity of cell to cell interaction for synthesis of ACTH receptors. In dispersed cells there is some mutual attachment of cells which normally have gap junctions. However, these would be eliminated by the usual physical methods of separation and therefore a comparison of the properties of unpurified and purified cells is relevant to this issue. It is also possible that the interstitial volume between the cells *in situ* serves as an intermediate zone for the transport of ions.[82]

It should also be emphasized that most of the theoretical advantages in the application of dispersed cells have yet to be proved rigorously. There is no doubt of the usefulness of separating dispersed cells to prepare homogeneous cell populations. However, the advantages in the improvement of the supply of nutrient and oxygen have not been firmly established. There is an impression in the field that dispersed cells respond to lower concentrations of stimulators than the corresponding tissue, but such increased sensitivity could be due to a lower basal rate of steroid production rather than improved access of humoral tropic factors and perhaps certain preincubation procedures could achieve this with adrenal tissue. Nevertheless the classical method of Saffran[83] involves preincubation of adrenal tissue and yet even then the threshold dose of ACTH required to stimulate steroidogenesis is much higher than for ZF cell preparations.

Even with these reservations, it can now be concluded, as it is hoped to demonstrate in the subsequent examples, that the aims of the use of dispersed cells have been largely realized and their application has at least usually coincided with an increased rate of progress in a particular scientific area.

B. INITIAL DISSECTION AND ENZYMIC DISPERSION

The use of isolated cells dispersed with enzyme, usually trypsin, for tissue culture has a long history. The extensive use of dispersed endocrine cells for acute experiments started about fifteen years ago at several centres and a view of its historical development probably depends critically on a reviewer's geographical position at the time. At the Worcester Foundation for Experimental Biology, the initial suggestion came from the fertile mind of O. Hechter and the idea was realized experimentally for adipose tissue at N.I.H. Bethesda by M. Rodbell and at Worcester by I. Halkerston for the adrenal cortex. Halkerston established that increased steroidogenesis after addition of coenzymes such as NADPH indicated damaged cells. He used preincubation with trypsin followed by bacterial collagenase to obtain optimum yields of undamaged cells.[84] Kono,[85] employing a variety of tissues, showed that a mixture of pure collagenases plus a protease was required for efficient dispersion of cells, and that the use of collagenase to prepare adipose cells was preferable to that of trypsin. Incubation of the dispersed cells was necessary to restore the activity of receptors after trypsin treatment.[86] Recently Neher[87] reported that the use of trypsin may render the cells more permeable to smaller molecules such as cyclic AMP.

After these pioneering studies, collagenase was used by Kloppenberg et al.[88] at Nashville to prepare mixed adrenocortical cells, probably mostly from the zona fasciculata. Haning,[89] in our group at Worcester, also used collagenase to prepare mainly ZG cells from rat adrenal capsules and later Williams[90] and Bing & Schulster[91] used essentially the same technique. Schulster was one of the first European workers to use dispersed adrenal cells and employed collagenase to prepare ZF-ZR cells from decapsulated rat adrenals.

Sayers[92] and Lowry et al.[93] used trypsin to prepare ZF-ZR cells from rat adrenals and Catt and co-workers (personal communication), in their earlier work, and later Edwards[94] used the same enzymic mixture to disperse zona glomerulosa tissue. In spite of the reports of Kono,[86] who studied other tissues, there is no firm evidence for loss of receptors from adrenal cells with trypsin. The advantage of using trypsin is that, following the incubation, its activity can be reduced by addition of specific inhibitors. Certain early studies suggested that the use of trypsin for cell dispersion resulted in greater sensitivity of ZG cells to angiotensin but this has not been substantiated. Although most workers use collagenase rather than trypsin, there is no good evidence to indicate a difference in the response of either ZF or ZG cells prepared by either procedure, so in subsequent discussion, the method of cell preparation will not be mentioned. Enzymic incubation, coupled with fairly simple mechanical agitation such as repeated pipetting, is usually sufficient to disperse adrenal tissue although Lowry et al.[93] and Schulster and co-workers[95] use more sophisticated methods involving rotating paddles or plastic pestles. It is generally not necessary to lower Ca^{2+} concentrations which has proved to be essential for the dispersion of pituitary cells[96] nor to perfuse the organ with enzyme solution as is necessary for cardiac preparations.[97]

The ZG cells from most species tend to be more difficult to disperse than ZF and ZR cells, which can be prepared with about equal ease. The ZF cells are more fragile, readily losing lipid droplets and this is very evident with guinea-pig adrenals.

If the cells are to be separated by the usual physical methods it is essential that the preparation is free of clumps since the yield depends critically on this. Halkerston,[84] in the earlier work, suggested the use of DNAase to avoid clumping. An optimum concentration of this enzyme and more thorough mechanical agitation may therefore be necessary if subsequent separation of cells is intended.

C. SEPARATION OF CELLS

Although specific affinity methods have played a role in the separation of cells of the immune system,[98] they have not been applied routinely to endocrine cells. However, general physical methods, if somewhat cumbersome and expensive, have proved to be useful. They usually depend on phenomena described by the well known Stokes' law:

$$\text{Terminal velocity on sedimentation} = K(\rho_C - \rho_L) \, r^2 . \mathbf{g}/\eta$$

where r and $\rho_C \equiv$ the radius and density of the cells and ρ_L and $\eta \equiv$ the density and viscosity of the suspending liquid. K is 2/9 for spherical cells. Two types of methods have been used, rate sedimentation (when $\rho_C > \rho_L$) and equilibrium density (when $\rho_C = \rho_L$)

Equilibrium density methods using a gradient in density of the solute and depending entirely on the density of the solute and suspended materials are the most commonly used for the separation of subcellular particles, but are more difficult to apply to whole cells. The required high density of the liquid is difficult to achieve without inconveniently increasing the viscosity or the general toxicity and osmotic effects of the solute, particularly because of contaminating salts and other impurities such as trace metals. Ficoll has been used, but this can yield solutions of high viscosity and its osmolarity is exponentially related to concentration.[99, 100] Metrizamide, an iodinated succinate, seemed a suitable solute but recent results indicate that it can markedly increase the density of several biological cells[100] including adrenal cells[101] (compared with ZF cells prepared using unit gravity sedimentation[75]) and Leydig cells[102] and reduce separation. This increase in density may be due to penetration or coating of the cell. Conn et al.[103] may even have damaged the cells by the use of metrizamide solutions (0–80%) of high osmolarity and so encouraged the entry of this solute into the cells.[99] Percoll, a silica suspension coated with polyvinyl pyrrolidone may provide the answer, as the densities of cells are unaltered after sedimentation with this solute, but it has been reported that enzymic activities are increased by this suspension.[104] Recently we have achieved satisfactory ACTH response of guinea-pig ZF and ZR cells separated by Percoll but further detailed work is needed on their morphological and functional properties. At the present time there is no well established solute for use in this method. In addition, for most of these solutes, if equilibrium is to be achieved in a reasonable time, higher centrifugal forces must be used and the effect of this on the cells is uncertain.[100] Nevertheless, Catt and co-workers[103] and Cooke and co-workers[105] found that Ficoll could be used to prepare relatively pure (about 60–70%) Leydig cells but the former workers[103] obtained even better results with metrizamide providing a 94% pure preparation of Leydig cells, 90% of which were viable as judged by their ability to exclude Trypan blue. This procedure using metrizamide (but not Ficoll) also eliminated red cells from the preparation. Janszen et al.[105] used a mixture of Ficoll (2–17%) and sodium metrizoate (4–6%) plus 0·2% albumin and obtained the relatively low densities for Leydig cells of about 1·07. Ficoll was found not to lower steroid production by the cells. It may be that the use of metrizoate in the mixture enables the use of lower Ficoll concentrations to achieve the required densities and the Ficoll plus albumin prevented the entry of the metrizoate into the cells. However, equal success has not yet been reported using a mixture of Ficoll and metrizoate for the separation of adrenocortical cells.

A mild procedure involving rate sedimentation with low concentrations of solute and low gravitational force, is the unit gravitational method which was established mainly by Miller & Phillips.[106] In this method, which depends as can be seen from the general equation, on both cell and solute density but mainly on radius when the density of the suspending solution is low, a concentration of albumin of 1–3% is usually employed (Fig. 3(a)). There is a gradient in solute concentration but this is to inhibit the effects of convection and streaming and is not fundamental to the method as it is with the equilibrium density approach. The sedimentation is carried out in a cold room without centrifugation and usually takes about 3 h. Detailed examination of the results for separation of Latex spheres by this procedure has shown that the

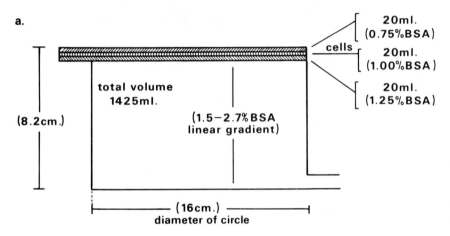

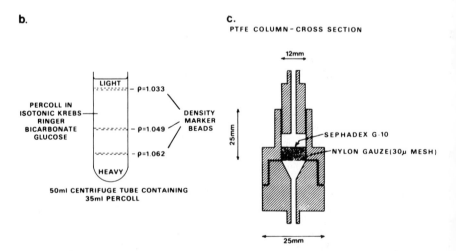

Fig. 3. Apparatus used for separation of cells. These are for (a) unit gravity sedimentation (b) Percoll equilibrium density and (c) column filtration methods.

resolution can be excellent and nearly theoretical.[75, 107] Purified adrenal cells obtained by this procedure are reasonably viable on morphological examination and in biochemical response using the normal stimulation ratios (stimulated/basal outputs) at higher doses as the criterion and we have been able to prepare ZG cells from rat adrenal capsules with less than 0·2% contamination by ZF cells.[75] Recently Braley & Williams[108] also reported purification of rat adrenal cells by unit gravity sedimentation, although the purity achieved for ZG cells (99·0%) was lower than that claimed by us, mainly due to high initial contamination with ZF cells. Maximal stimulation ratios were not calculated by Braley & Williams as they appear to regard the necessarily higher doses used by us as "pharmacological". However their data show that the maximal stimulation ratios are not changed by unit gravity sedimentation unless ZF cells, which are removed, respond to the stimuli employed. They also confirm that the basal production of corticosterone and aldosterone was about halved after sedimentation,[75, 108] although the threshold sensitivity to the specific ZG cell stimulators, angiotensin II and K^+, was not affected.

Starting with rat decapsulated adrenals, ZF and ZR cells have been prepared by unit gravity sedimentation in the same system as used for ZG and ZR cell separation with less than 10% cross contamination.[13] These separations are with cells which differ in both radius and density which, because of related factors, often tend to act in opposite directions to reduce separation. However, with Coulter counter control, adequate preparations can be obtained routinely. Recently, also in our laboratory, for the separation of guinea-pig ZR and ZF cells which are nearly equal in size range (median diameter about 20 μm), the concentration of albumin has been increased by a factor of 2·5 so increasing the density of the liquid and decreasing the differences between cell and solute densities.[109] The method is then operating more towards the conditions of equilibrium density without the undesirable very high concentrations of albumin and centrifugal force. However the recent results of Dr Hyatt in our laboratory indicate that the Percoll method may be more convenient (Fig. 3(b)). One disadvantage of the unit gravity sedimentation method generally, as shown in the apparatus of Fig. 3(a), and also of the equilibrium density approach in the same apparatus for whole cells is that large volumes of solute must be used to avoid the phenomena of streaming and aggregation which destroy resolution[75] and this is inconvenient when recovering the cells and expensive with the high cost of albumin.

A method which avoids the use of these large volumes of albumin solution and which depends on size only, unlike the other two approaches, is desirable. Recently this has been achieved by McDougall & Williams,[110, 111] with the same type of gel column employed by Lowry & McMartin[76] for superfusion experiments, but using Sephadex and modified in design to reduce dead volume effects for use in Ca^{2+} washout procedures[110] to be described later (Fig. 3(c)) (Section III.A.4 and III.B.6). There is a differential passage of cells through such columns depending on the radius of the cells and on the mesh size of the Sephadex. The use of a suitable mesh size (G-10) allows the passage of rat ZR cells (median radius 8 μm) but not of ZF cells (median radius 18 μm) in the decapsulated gland preparations. The use of mesh size G-15 allows the passage of ZG cells (median radius 9 μm) but not of ZF cells from the capsular gland preparation. Separation seems to be on a size basis only, although judging from the ease with which erythrocytes pass, flexibility of the cells and their effective radius on the column is probably a consideration.[111]

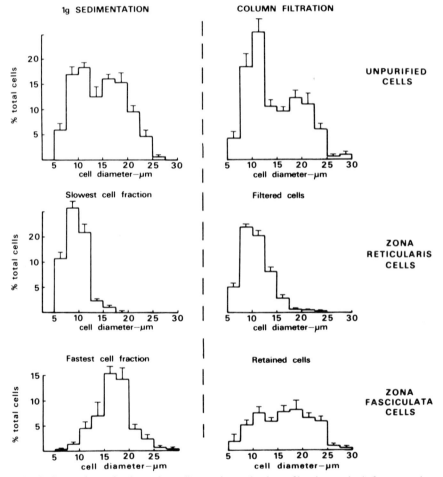

Fig. 4. Comparison of unit gravity sedimentation and column filtration methods for separation of rat ZR and ZF cells. Histograms of cell sizes, unpurified and purified.

As the smaller cells, the ZG and ZR cells, are usually the ones required in purified form and are more conveniently prepared, this is an appropriate, rapid and inexpensive method for adrenal cells. Their morphological and response characteristics are similar to cells prepared by the higher resolution, unit gravity sedimentation method although more debris and erythrocytes are present together with the smaller cells (Fig. 4). Recently Catt and co-workers[112] have used the McDougall–Williams[110, 111] method to obtain pure ZG cells for studies with ACTH stimulation when homogeneous cell preparations are essential. Only for special purposes, when it is desirable to eliminate contaminating non-steroidogenic cells and debris, e.g. in the measurement of cellular K^+ concentration, is the more expensive method now necessary.

There are now three satisfactory physical methods for the separation of cells: (i) unit gravity sedimentation depending on the [diam.]^2 and density of the cells; (ii) column filtration depending on the [diam.]^2; and (iii) Percoll equilibrium density depending on density only (Fig. 3). Therefore, one of these can be chosen to solve most separation problems. Even the most difficult task such as the purification of pituitary cells should be successfully accomplished by a combination of two of these methods.

III. Studies with Dispersed Cells

A. ZONA FASCICULATA (ZF) CELLS

(1) Specificity of response

Using dispersed adrenal cells, particularly those made from decapsulated rat adrenal tissue which consist of a mixture of ZF and ZR cells, the peptide structure–activity criteria for eliciting steroidogenic response have been studied.[113, 114] These results show that the ZF cells are quite specific in response and only closely related analogues of ACTH are effective stimuli. Serotonin and K^+ do not stimulate ZF cells as they do the ZG cells.[89] Recent results show that β-endorphin stimulates corticosterone production in dispersed rat ZF cells,[115] but the potency is 1000 times less than that of ACTH and the maximum steroid production is lower.

It remains to be seen whether the impurity in the preparation of [Asn^1] angiotensin II (Hypertensin) which stimulates rat ZF cells[91, 116] (see Section III.B.1) is a new type of stimulatory substance or whether it is due to contamination with ACTH itself. Whatever the reason for this activity, Tait et al.[116] found that pure [Asn^1] and [Asp^1] angiotensin II, even at concentrations of 10^{-4} M, do not stimulate corticosterone production by rat ZF cells.

However, Peytremann working in the laboratories of Liddle[117] and Vallotton[118] found that both angiotensin II and [des-Asp¹] angiotensin II stimulate steroidogenesis by bovine ZF cells and the amount required for threshold stimulation of both ZF and ZG cells is similar. In earlier work, Kaplan & Bartter[119] also found that angiotensin II stimulated bovine zona fasciculata tissue.

It is tempting to explain these results by the effect of an impurity in the Hypertensin preparation used as angiotensin II in some of these experiments. However, the activity of the [des-Asp¹] angiotensin II, which was equipotent with the Hypertensin preparation of angiotensin II on the ZF cells, and the effectiveness of an antagonist [Sar¹Ala⁸] angiotensin II in the studies of Peytremann[118] does not fit this idea. Also recently Bravo et al.[120] reported marked effects of pure angiotensin II in vitro on steroidogenesis in cat ZF cells as did Catt[112] in canine ZF cells. There are reports of stimulation of cortisol in hypophysectomized dogs and man in vivo by high doses of angiotensin II (Hypertensin).[121–124] Oelkers[38] has compared the effects of Hypertensin and of [Asp¹Ile⁵] angiotensin II in man after infusion of physiological amounts of peptide which markedly stimulated aldosterone. Neither cortisol nor other 11β-hydroxy steroids were stimulated by either preparation. This lack of response of cortisol, however, could be due to the operation of an ACTH feedback system in human subjects with intact pituitaries.

There have been recent suggestions that natural angiotensin II at high concentrations may stimulate the zona fasciculata in man, although with abnormal adrenal tissue.[125] In those with the adrenogenital syndrome involving 21-hydroxylase deficiency, after cortisol treatment plasma ACTH levels can be normal with high renin and angiotensin II concentrations. In these circumstances levels of serum 17α-hydroxyprogesterone, which presumably originates in the zona fasciculata, are also raised. Leutscher explained this as the result of the zona fasciculata stimulation by the high levels of angiotensin II. However, according to quantitative considerations the increase in 17α-hydroxyprogesterone may simply be due to the block in subsequent conversions to cortisol and 11-deoxycortisol. Also McKenna (personal communication) has reported that if fluorohydrocortisone (a steroid with both mineralocorticoid and glucocorticoid activity) is used in therapy to correct both angiotensin and ACTH levels then 17α-hydroxyprogesterone is still increased. There is therefore clearly a need for further work in vitro and in vivo on the ZF cells of various species using pure natural angiotensin II before this important question can be decided. As there is also the possibility of zona glomerulosa contamination of the zona fasciculata preparations explaining some of these in vitro results, as the maximum steroidogenic ratio was low and the basal aldosterone production was high,[126] it will be desirable to use homogeneous ZF cells.

ZF cells are stimulated strongly and specifically by ACTH and its close analogues. Endorphin peptides may stimulate weakly. High doses of pure angiotensin II may stimulate the zona fasciculata of some species. However, many but not all such effects could be due to an ACTH-like contaminant in a commonly used preparation of the peptide.

(2) Receptors

Using ACTH diazotized to polyacrylamide, Richardson & Schulster[95] showed that the steroid output of dispersed ZF-ZR cells could be stimulated, an effect unlikely to be due to soluble active peptide fragments as was suggested for similar experiments with insulin.[127] It appears that ACTH binds to the plasma membrane, adenylate cyclase is also associated with the same membrane, and cyclic AMP acts as the mobile messenger within the cell. Various studies either directly, using labelled ACTH,[128] or indirectly with ACTH analogues,[129] have indicated that there are at least two types of receptor for this peptide hormone. One has low affinity and the other has an association constant some 40-fold higher with fewer sites, although the results could be interpreted in terms of heterogeneously labelled hormone, little of which might behave as the natural peptide and/or the binding by a mixed population of cells, particularly from both the zona fasciculata and reticularis. Preliminary results show that purified ZR cells bind ACTH less strongly than homogeneous ZF cells.[130]

Probably because of the lack of availability of a suitably labelled hormone, there have been no published studies on the effect of administration of ACTH on the affinity and number of its receptor sites on adrenal cells. However, studies on other endocrine cells have shown that one of the effects of a tropic hormone is to alter the number of its own receptor sites; reduced numbers of sites when the hormone concentration is increased above normal levels and vice versa. This phenomenon has been termed "down-regulation" and results in overall negative feedback control.[131] In a few situations, the affinity of the receptors is also lowered as a result of tropic hormone administration by mechanisms of negative co-operativity which have not yet been elucidated in detail. Other than studies on the zona glomerulosa,[132] the only reported example of the opposite effect of an increase in the number of receptor sites with tropic hormone administration ("up-regulation" with overall positive feedback control) is the effect of prolactin on liver cells.[133]

Because of the similarity of the characteristics of the proteins involved, it was originally thought that the receptor and adenylate cyclase units might be identical or intimately complexed. Recently, for a variety of cells, they have been separated. Schulster *et al.*[134] transferred β-adrenergic receptors from turkey erythrocytes to cultured erythroleukaemic cells or to mouse adrenal tumour cells responsive to ACTH. The recipient cells responded to isoprotenerol with an increase in intracellular cyclic AMP. Ovarian LH receptors have also been transferred functionally to adrenal cells containing adenylate cyclase.[135] It may be that the hormone receptor and adenylate cyclase units float in a fluid membrane and are brought together by the action of the tropic hormone.[131]

If there are several forms of second messenger control, e.g. Ca^{2+} and cyclic AMP (Sections III.A.3 & 4), it might be that they have several types of receptor each coupled to one control mechanism. Schulster[129] has recently suggested, as a result of studies with different ACTH analogues, that there are two types of receptor in ZF cells, one coupled to the adenylate cyclase system and another whose function is unknown but may be coupled to Ca^{2+} metabolism.

There may be at least two types of receptor for ACTH in ZF cells perhaps coupled to different factors. The binding receptor and adenylate cyclase complexes are closely connected but can be separated in different cellular preparations.

(3) Cyclic AMP

Several early studies with dispersed adrenal cells clearly showed dissociation between cyclic AMP and steroid outputs with ACTH stimulation of ZF–ZR cells.[136, 137] This was particularly evident at higher outputs when raised nucleotide values still accompanied increasing ACTH stimulation after maximum steroid production had been achieved. As for similar results with adrenal tissue, this supports the concept of spare receptors.[48, 128] More critical for the validity of the Sutherland–Garren hypothesis[41, 45] with whole cells and in the lower range of values, highly significant increases in steroid output occurred without corresponding significant rises in cyclic AMP. However, in one detailed study,[138] which is usually quoted as showing this effect, the authors found this dissociation at only one dose of ACTH and considered that their overall results showed good correlations and supported the cyclic AMP messenger hypothesis. A rather similar reasonable correlation is shown between cyclic AMP and steroidogenesis at low ACTH concentrations (about 10^{-11} M $^{1-24}$ACTH) in one study with human adrenal dispersed cells,[139] although in another study there was a dissociation.[140] Dichotomy in the lower range of stimulation is clearly shown in other studies on adrenal cells and also in a variety of different endocrine cells.[141] Also certain dynamic studies both in an *in vitro* superfusion adrenal cell system[142] and *in vivo* in sheep with exteriorized cervical adrenal transplants,[143] in which the output of cyclic AMP and cortisol (and aldosterone) was compared, do not support an obligatory role for cyclic AMP in ACTH action.

Recently workers at N.I.H., Bethesda and in Neher's laboratory in Basle have succeeded in obtaining good correlations of cyclic AMP and steroidogenesis in testicular[144] and adrenal cells[137, 145] after systematically altering the incubation conditions. Certain of the early reports from the Bethesda group claimed that the factor in this success was the measurement of cyclic AMP bound to a receptor protein rather than the extracellular nucleotide.[137, 145] However, detailed examination of the two major papers reveals that many factors contribute to the better correlation. The measurement of bound cyclic AMP may be a minor factor, particularly if phosphodiesterase inhibitor is added to the incubations before extracellular cyclic AMP is measured. Sala et al.[145] report that extracellular cyclic AMP was significantly increased at ACTH concentrations of 10^{-12} M ($p < 0.05$), 10^{-11} M ($p < 0.05$) and 10^{-10} M and above ($p < 0.01$) whereas bound cyclic AMP was increased at 10^{-12} M ($p < 0.05$) and 10^{-11} M ($p < 0.01$). The significance of the effect at 10^{-10} M is not given for the rise in bound nucleotide, but its fractional increase is clearly less than for the extracellular nucleotide. In the paper by Podesta et al.,[137] appropriate direct comparisons are not reported but it seems that 10^{-12} M ACTH caused a marginally significant rise in extracellular and total intracellular cyclic AMP in incubations treated with phosphodiesterase inhibitor, 3-isobutyl-1-methylxanthine (MIX), and an equivalent increase in bound cyclic AMP with or without the addition of the phosphodiesterase inhibitor, which has little effect on the bound values. The overall advantage of measuring bound rather than extracellular (with MIX added) cyclic AMP for testis and adrenal cells to achieve good correlations of nucleotide and steroid values, except perhaps conceptually, is rather doubtful, as was clearly stated by Catt and co-workers.[145] The presentation of results in terms of occupancy level of receptors rather than extracellular cyclic AMP appears to give the impression of better correlations.[137, 145] However, and this applies to other comparisons of the correlation of aspects of nucleotide values and steroidogenesis, the judgement must obviously be based on comparison of the variance or variance/slope rather than only the slope of the plot of the relevant variables.

Other factors which are probably more important are:

(i) The use of a phosphodiesterase inhibitor, e.g. MIX, at a critical concentration,[137] which does not inhibit steroidogenesis nonspecifically as in earlier studies with theophylline.[146] It should be emphasized that there is nothing fundamental in this effect. In general, the higher the rate of degradation of a substance the more rapidly and closely its level follows its production rate. Therefore better correlations after the use of a phosphodiesterase inhibitor indicate that the nucleotide values have higher stimulation/basal ratios because of a significant blank either due to methodological difficulties or because there is a significant amount of stable cyclic AMP which is less stimulated or degraded. It should also be noted that nonspecific effects of phosphodiesterase inhibitors could be due not only to the direct inhibition of steroid production as seen with theophylline for ZF cells[146] and even with MIX for ZG cells[112] (Section III.B.4) but also because of the possibility that methylxanthines may directly affect the intracellular transport of Ca^{2+}.[147]

(ii) The use of more appropriate and shorter time intervals (usually 1 h) following kinetic studies[145] which showed the bound and intracellular nucleotide output to be increased maximally after about 15 min, and extracellular nucleotide after 60 min. However, this is probably not an important effect unless the phosphodiesterase activity were to be very high, as accumulated, not peak, extracellular values are usually measured.

(iii) Preincubation of the cells for 2 h before stimulation to provide more constant and lower basal conditions. Data are not provided by Catt and co-workers to enable the effect of this procedure to be judged, but it could be quite important; it was first shown to be a valuable procedure by Cooke et al.[148] using interstitial cells and was also employed by Saez et al.[140] with human adrenal cells.

(iv) The use of the specific and sensitive assay of Harper & Brooker[149] based on the acetylated nucleotide which not only allowed the measurement of bound cyclic AMP but also gave lower basal production rates of any nucleotide component.

(v) In the case of Leydig cells, Catt and co-workers used cells purified by metrizamide density gradient sedimentation as previously described[104] (Section II.C) and obtained a more sensitive and maximum production of both steroid and nucleotide output with HCG stimulation for a particular cellular fraction. Although, as previously discussed, there are certain problems in such methods of separation using metrizamide[102] the authors claim that a preparation of 94% Leydig cells can be obtained which respond more sensitively than the unpurified cells.[104] The advantages in terms of correlation of nucleotide and steroid values are again small, but significant. In the case of adrenal cells, a mixture of ZF and ZR cells has usually been used and this could result in poor correlations of nucleotide and steroid outputs. Hyatt et al.[150] in our laboratories have shown that when adrenal cells are separated by the high-resolution method of unit gravity sedimentation, the purest fraction of the preparation of ZF cells gives a greater (maximal) response of cyclic AMP to ACTH than unpurified adrenal cells. ZR cells give a much lower nucleotide response. As these cells contribute significantly to the total nucleotide output of unpurified cells but not to the overall corticosterone values, it seems that the use of purified ZF cells could improve the correlations of cyclic AMP and steroid levels.

With these modifications to the earlier experimental procedures, all of which make small but significant contributions to the improvement in correlations, the Sutherland hypothesis can now be amply confirmed by this approach in ZF cells. The only results of this kind which do not fully fit the hypothesis are those with certain fragments or derivatives of ACTH, e.g. [5-24]ACTH and the o-nitrophenyl sulphenyl (NPS) derivative[113, 129, 151, 152] which stimulate steroidogenesis but have little effect on nucleotide output. Although these have not been investigated in the modified systems of Catt and Neher and co-workers, this may indicate the presence of receptors not coupled to adenylate cyclase as previously discussed (Section III.2).

Similarly, early results of Richardson & Schulster[153] using adrenal cells after ACTH, and Catt and co-workers[154] using interstitial cells after gonadotropin stimulation showed dissociation of measured cyclic AMP-dependent kinase and steroid output with weak activation by tropic hormone. However, later results showed good correlation of cyclic AMP-dependent kinase and steroid levels for both types of cell.[144, 155] This was due to modifications of the conditions such as preincubation as employed for human cells by Saez et al.[140] and probably more importantly, the isolation of specific phosphorylated proteins responsive to cyclic AMP levels. Therefore, for ZF cells correlation of cyclic AMP bound to the regulatory subunit of the kinase and the appropriate cytoplasmic protein, labelled after incubation with ^{32}P, with the steroid levels[155] now satisfactorily confirms the Sutherland theory[41] as modified by Garren and other workers.[45]

During the period when a dichotomy between cyclic AMP and steroid outputs was evident, Sharma and co-workers[156] put forward the theory, based on substantial experimental evidence, that in adrenal cells, cyclic GMP may be the dominant messenger at lower levels of ACTH stimulation, whereas cyclic AMP may be more important at higher steroid outputs. This is a separate question from whether there is participation of guanyl nucleotides (GTP) in the adenylate cyclase system which is likely.[157] However, as regards the role of cyclic GMP as a separate messenger, in the paper by Catt and co-workers[145] which showed a good correlation of cyclic AMP (extracellular, intracellular and kinase bound) and steroid output from ZF-ZR cells, it was stated that there was dissociation of the cyclic GMP and steroid levels with the response to this nucleotide being confined to the extracellular compartment. Also very high concentrations of cyclic GMP are required to stimulate steroidogenesis although this may be a matter of poor transport of the exogenous nucleotide. A recent paper of Sharma[158] presents data which maintain his original position and stresses the essential role of Ca^{2+} in the cyclic GMP response but further work is needed to examine the function of the response of the cyclic GMP in the extracellular compartment. It may be that with the elucidation of a possible Ca^{2+}-non-cyclic AMP mechanism[129] for the control of steroidogenesis, a role of cyclic GMP will be recognized as it has been in the corresponding inositol phospholipid systems.[159]

With the usual experimental conditions with dispersed cells, there is a discrepancy between the steroid and cyclic AMP outputs after weak ACTH stimulation. However, using preincubation, a phosphodiesterase inhibitor (or measurements of bound cyclic AMP), sensitive and specific nucleotide assays, and shorter times of incubation, good correlations of steroid and nucleotide levels can be obtained. The use of purified ZF cells, with the usual contaminating ZR cells eliminated, should lead to even better correlations. Similarly cyclic AMP-dependent protein kinase activity and steroidogenesis were found to be dissociated but with modification of the incubation conditions and more specific protein measurements, these can now be correlated. It has been suggested that a relevant kinase is controlled by cyclic GMP particularly at very low ACTH concentrations but this is not as yet generally accepted. The major data remaining to be explained before the Sutherland theory is completely accepted as being the sole mechanism applicable to ZF cells have been obtained with certain fragments and derivatives of ACTH which stimulate steroidogenesis maximally but have only a marginal effect on cyclic AMP.

(4) Calcium

Many experiments on both ZF[160] and ZG[161] cells have been performed in which extracellular Ca^{2+} and also, by the action of chelating agents, intracellular Ca^{2+} is lowered and basal and stimulated steroidogenesis is observed. Ca^{2+} transport inhibitors[161, 162] have also been used. These, in principle, show that:

(i) If steroidogenesis is unaffected, normal Ca^{2+} concentrations are not required for full demonstration of steroid outputs and therefore also changes in Ca^{2+} transport with stimulation are unlikely.

(ii) If steroidogenesis is reduced then normal Ca^{2+} concentrations are required for full basal or stimulated steroid output (permissive effects of Ca^{2+}) but it cannot be concluded from this type of observation that changes in Ca^{2+} transport are involved in stimulation (non-permissive effects). The latter effect must be shown by more direct observations.

Direct studies on Ca^{2+} metabolism of cells, with and without stimulation, would ideally reveal effects on the influx and efflux permeability properties of the plasma membrane and intracellular transport of the divalent ion. Experimentally, these can be shown by effects on the uptake of isotopic Ca^{2+} or on the efflux of the labelled ion from preloaded cells.[163] In practice, the interpretation of effects on uptake, although the only type likely to reveal rapid events, are difficult to dissect as usually the contribution of extracellular ion must be eliminated by the passage of the cells through oil of suitable density before the measurement of their ionic content.[164] This, however, makes such investigations so tedious that various doses of the stimulatory factors cannot be used and appropriate correlations established. On the other hand, efflux curves, after prelabelling of the cells with radioactive Ca^{2+}, can be readily investigated after a certain time interval and the effects of the extracellular ion eliminated by their analysis. Such curves have been analysed in detail in a number of situations by Borle.[165] It emerges that effects on these decay times can be due to a combination of changes in the specific transport properties of the plasma membrane or in the intracellular movement of Ca^{2+}. However, changes in washout decay times in endocrine systems are thought to be nearly always due to alterations in intracellular Ca^{2+} concentrations.[166, 167]

Therefore, changes in the $^{45}Ca^{2+}$ washout decay characteristics of adrenal cells after about 20 min are the most reliable both intrinsically and because of the possibility of obtaining enough data.[110] It would be most fortunate, therefore, if changes in the transport of the ion brought about by a tropic factor had effects on these decay times. Nevertheless, it should be noted that alterations in other aspects of the transport of the ion, such as very rapid changes, could be more directly connected with the mechanism of action of the factor and yet these could not be readily observed with present methods.

The role of Ca^{2+} in the steroidogenesis of the zona fasciculata-reticularis of the adrenal cortex was reviewed in 1975 by Halkerston[160] and most of his conclusions still stand. Lefkowitz et al.[54] (see Section 1.D) showed that concentrations of more than 1 mM Ca^{2+} inhibited binding of ACTH to receptor sites and activation of adenylate cyclase in subcellular membrane particles from mouse adrenal tumour tissue. However, Sayers et al.,[168] using adrenal dispersed cells, found that in the range 0–7·65 mM Ca^{2+} the response of corticosterone to ACTH was increased both as regards maximum output and threshold sensitivity. Again, using whole cells, two other groups[169, 170] obtained similar results in that, in the absence of Ca^{2+} in the incubation medium, ACTH did not stimulate steroid output from dispersed cells. Added cyclic AMP under these conditions elicited a positive but reduced steroidogenic response.[169, 170] Again there were conflicting results between experiments with whole cells or subcellular particles as concentrations of Ca^{2+} in the range used to stimulate steroidogenesis with whole adrenal cells inhibited adenylate cyclase activity in subcellular particles from adrenocortical tissue.[55, 171, 172] It seems, as suggested by Sayers et al.,[168] that the internal ionic environment of whole cells is not so affected by changes in extracellular Ca^{2+} as is that of subcellular particles. If this suggestion is of general relevance caution is needed in interpreting findings with adrenal mitochondria. Ca^{2+} (10^{-4} M) increased high-spin ferric cytochrome P-450 bound to cholesterol, an effect also brought about by ACTH.[173] These reservations also apply to studies by Farese[174] who demonstrated an effect of Ca^{2+} on protein synthesis in microsomes and soluble cell fractions of non-ACTH stimulated adrenals although in these latter experiments, concentrations as low as 10^{-7} M Ca^{2+} were effective.

Although studies with whole adrenal cells may be more relevant to physiological conditions, it is more difficult to investigate the effects of Ca^{2+} on various points of action of ACTH with this type of preparation. Nevertheless, it is to be expected that effects on the cytochrome P-450 system in dispersed whole cells could be studied both by optical and electron paramagnetic resonance methods with the increased sensitivity of modern equipment.[173] Direct studies on the effects of Ca^{2+} on labelled ACTH binding to whole cells have not been reported, probably again due to the lack of a suitable preparation of radioactive peptide.

The experiments so far described demonstrate only the permissive requirements for Ca^{2+}. Unfortunately, due to technical difficulties, direct *in vitro* experiments of the kind found to be revealing for other tissues, such as the study of $^{45}Ca^{2+}$ washout curves, have not been common. Leier & Jungmann[175] studied the uptake of $^{45}Ca^{2+}$ and measured the intracellular nonisotopic Ca^{2+} concentrations of adrenals. [^{3}H]Inulin was used to correct for extracellular ion. Whole adrenals with the complete mixture of cells were analysed. They found that ACTH amd cyclic AMP (and theophylline) increased the $^{45}Ca^{2+}$ uptake, 90 to 180 min after hormone addition. Inhibitors such as elipten and cycloheximide eliminated both $^{45}Ca^{2+}$ uptake and increased steroidogenesis after ACTH stimulation. Jaanus & Rubin[176] however report no increased influx of Ca^{2+} with ACTH. The studies of Mathews & Saffran[177] on rabbit adrenal glands emphasize the dissociation of membrane polarization and the secretion of corticosteroid. They emphasize that Ca^{2+} deprivation has comparatively little immediate effect on the action of ACTH on the adrenal cortex but rapidly inhibits the stimulated secretion of the adrenal medulla and other cells which store the secretory product; lanthanum and local anaesthetics, such as nupercaine, inhibited steroidogenesis. The experiments of Rubin et al.[178] with the perfused cat adrenal present convincing evidence that Ca^{2+} may play a dominant role in the release of steroid in this preparation where protein and steroid output can be correlated. There may be storage of steroid product in electron dense organelles of the cat adrenal[179] and the release would then be expected to be strongly influenced by Ca^{2+}. Similar electron dense granules have been observed by Nussdorfer et al.[180] in rat adrenals after treatment with vinblastine and this was correlated with an increase in intracellular corticosterone, whereas the plasma levels were reduced. However, Ray & Strott[181] suggested that vinblastine and other antimicrotubular drugs stimulate steroidogenesis by disruption of the microtubules resulting in preparations equivalent to cell-free homogenates. Although Vinson and co-workers[182] have also indicated that some stored steroid may be released with protein from rat adrenals, the general applicability of work with the perfused cat adrenal remains to be established. Nevertheless the work of Shima,[161] who showed that lanthanum inhibited ACTH effects in the rat adrenal also has to be explained, although the interpretation of results with such inhibitors must be viewed with caution[162, 183] (Section III.B.6).

Washout curves obtained after prelabelling superfused ZF rat adrenal cells with $^{45}Ca^{2+}$, (Sections II.C and III.B.6), show no effect with 3×10^{-8} M ACTH, a dose which increases steroidogenesis. Angiotensin II increases the efflux of $^{45}Ca^{2+}$ from ZG cells using the same procedure. Unfortunately, only one dose of ACTH was used.[77] Rasmussen in a review shows effects of ACTH on Ca^{2+} efflux[167] but these results cannot as yet be confirmed (Rasmussen, personal communication). Clearly the results of direct studies (either with measurement of nonisotopic or isotopic Ca^{2+}) are incomplete and further work is required before any reliable conclusions can be drawn. The use of Ca^{2+} ionophores will be useful but at the present time the specificity of their effects remains to be established.

Neher and co-workers[184, 185] have gone so far as to claim that Ca^{2+} is the most important second messenger in the action of ACTH on ZF cells with cyclic AMP playing a permissive role, a view opposed to current opinion. Cells isolated in a low Ca^{2+} medium with the use of trypsin, were permeable to Ca^{2+} and a suspension of colloidal Ca^{2+} raised steroidogenesis without the addition of corticotropin. It remains to be seen whether this is the form of Ca^{2+} normally present in the cytoplasm for it may aggregate enzymic systems and raise steroidogenesis nonphysiologically. Nevertheless, recent results of Podesta et al.[183] have emphasized that this effect of colloidal Ca^{2+} acts by increasing cyclic AMP levels. Some transport of Ca^{2+} may be necessary for the action of ACTH as shown by a Ca^{2+} transport inhibitor, verapamil. However, this could not be demonstrated by $^{45}Ca^{2+}$ studies or by the action of Ca^{2+} ionophores.[186]

Rubin & Laychock[187] also suggested, from preliminary studies, that Ca^{2+} is involved in the ACTH stimulation of prostaglandins, probably by an effect on phospholipase A_2. As will be discussed (see Section III.B.1), there is evidence, although rather confusing, that prostaglandins may have a direct effect on steroidogenesis in the adrenal cortex.[188] Lowering of Ca^{2+} concentrations in vitro inhibits both prostaglandin and steroid production in response to ACTH.[187] They also suggest that when the NPS-ACTH derivative increases steroidogenesis without raised cyclic AMP levels it may act through the Ca^{2+}-prostaglandin-activated mechanism via phospholipase A_2,[187] although the latter process in other situations may help to increase cyclic AMP. This is an interesting theory which at least explains in some detail the otherwise mysterious Ca^{2+}-non-cyclic AMP mechanism and further critical experimental evidence on these all embracing views is awaited with interest.

Ca^{2+} is required permissively for full demonstration of ACTH stimulation of steroidogenesis. The major site of action is in the transfer of the information from the binding receptor to the adenylate cyclase complex although there may *be subsequent requirements beyond the production of cyclic AMP. Participation of Ca^{2+} in the actual process of binding has not yet been proved. The role of non-permissive effects of Ca^{2+} (changes in Ca^{2+} transport) on ACTH-cyclic AMP-stimulated steroidogenesis is not yet established. The main evidence relies on indirect studies with Ca^{2+} transport inhibitors. Suggestions for Ca^{2+} as an independent second messenger rest on: (i) the existence of two sorts of receptor for ACTH, one concerned with cyclic AMP and the other possibly with Ca^{2+} transport. The evidence for the two receptors is mainly from studies with ACTH fragments and NPS-ACTH derivatives which are steroidogenic but do not markedly stimulate cyclic AMP; (ii) steroidogenic (without ACTH) stimulatory properties of colloidal Ca^{2+} although a recent report does not support the existence of a Ca^{2+}-non-cyclic AMP mechanism from such data; (iii) a Ca^{2+}-prostaglandin-non-cyclic AMP mechanism involved after stimulation (or release) of steroidogenesis in ZF cells by NPS-ACTH in adrenal glands which store steroids.*

There is increasing evidence for both release and biosynthetically active mechanisms being involved in hormone output by the adrenal cortex. Perhaps much confusion could be avoided in the future by specific examination and reference to the role of Ca^{2+} in these two distinct processes.

B. ZONA GLOMERULOSA (ZG) CELLS

(1) Specificity of response

For over ten years we and our co-workers have studied the complex nature of the response of dispersed rat capsular adrenal cells to various stimuli. The responses[116] of the aldosterone and corticosterone outputs of these preparations (mainly ZG cells with about 5% ZF cell contamination) to various stimuli are of four types (Fig. 5):

(i) As shown by the corresponding tissue studies of Muller,[58] variations in K^+ concentration gives a maximum aldosterone response at 5·9–8·4 mM K^+, about 6-fold higher than the control output at 3·6 mM. At higher concentrations, such as 13 mM, the response decreases.

(ii) Again as shown in Muller's work,[58] serotonin (at a concentration of 10^{-4} M) gives only a slightly lower maximal aldosterone response than does K^+ but this does not decrease significantly at higher concentrations. Serotonin gives a significant response at about 10^{-9} M.

(iii) As was first shown by Williams et al.,[90] Catt et al.[189] and Bing & Schulster,[91] [Asp^1 Val^5] angiotensin II gives a significant steroidogenic response at low concentrations corresponding to physiological concentrations in peripheral plasma. More recently we have found[116] that this angiotensin, the natural peptide in cattle and sheep, gives a significant response at about 10^{-10} M and a constant maximal response from $2 \cdot 5 \times 10^{-8}$ M. This maximum response is however only about one-half that found with serotonin or K^+ as stimuli. [Ile^5] angiotensin II, the natural peptide[190] in the human, horse, rat and hog, has similar activity. Using a preparation of [$Asn^1 \, Val^5$] angiotensin II (Ciba Hypertensin), the initial maximum response, which is similar to that seen with the natural [$Asp^1 \, Val^5$] angiotensin II, is increased at a higher concentration (from $2 \cdot 5 \times 10^{-5}$ M) and may eventually be greater than the maximum response with K^+. This additional response is, to a major extent, due to stimulation of the contaminating ZF cells and is not seen, as was first reported by Bing & Schulster,[91] with high concentrations of the pure [$Asp^1 \, Val^5$] angiotensin II. It was also not seen with pure [$Asn^1 \, Val^5$] angiotensin II (the natural angiotensin of the Japanese goose fish)[190] which has a single maximum plateau of response of about the same magnitude as with [$Asp^1 \, Val^5$] angiotensin II. The additional stimulation could be due to an impurity in the Hypertensin and this may have caused considerable confusion for both *in vitro* (previously discussed in Section III.A.1) and many, but not all, *in vivo* studies which have indicated stimulation of zona fasciculata steroidogenesis by angiotensin II (Section III,A.1).

(iv) ACTH (Ciba Synacthen) at 3×10^{-11} M gives a significant steroidogenic response with rat capsular cells. Higher concentrations (3×10^{-10} M to $7 \cdot 5 \times 10^{-9}$ M), however, give no constant maximum and the response can be much greater than for any other stimulus. This additional response in rat cells is again largely due to the stimulation of steroid precursors, such as deoxycorticosterone and corticosterone, originating from the contaminating ZF cells[116] although Peytremann *et al.*[191] claimed that this enhanced ACTH response is not explained in this manner for bovine ZG cells. Purified ZG cells show about an equal maximum response to ACTH as to K^+ and serotonin and also to high concentrations of Hypertensin.[75] This may indicate that the steroidogenic response for all stimuli (except to pure angiotensin II which is lower) is similar for homogeneous ZG cells. On the other hand, it could be that there are sufficient contaminating ZF cells remaining even after unit gravity sedimentation (only about 1 in 2000 would be sufficient) for an additional small response to ACTH (and to the contaminating material in Hypertensin which also has a high steroidogenic stimulation ratio in ZF cells), whereas the response of completely non-contaminated ZG cells could be similar to that of pure angiotensin II. It may even be that the answer to this question requires the use of single cells. We[116] found that the threshold sensitivity for capsular cells with ACTH stimulation (3×10^{-11} M) is lower than for ZF cells (3×10^{-12} M).

Several groups have pursued similar studies with rat capsular adrenal cells and there are now only a few discrepant results. Braley & Williams[192] have found equal threshold sensitivities for ACTH (4×10^{-12} M) in ZF and ZG cells because of apparently increased sensitivity of the ZG cells and suggested that this corresponds to the results of human studies *in vivo*. However, quite apart from species differences, *in vivo*, ACTH reaches the ZG cells before the ZF cells and it seems possible that this discrepancy was due to a higher ZF cell contamination in the Braley & Williams preparation. This has already been confirmed in a recent paper by these authors[108] which reports that, when their ZG cells are purified by unit gravity sedimentation, their threshold sensitivity to ACTH decreases tenfold (see Section II.C).

The reported threshold sensitivities of the steroidogenic responses of the various preparations of rat capsular adrenal cells to angiotensin II are similar for different groups. The threshold dose for the preparation by Catt and co-workers[193] of canine ZG cells for aldosterone output is slightly lower (10^{-11} M angiotensin II) whereas the ACTH threshold is higher (10^{-10} M Synacthen) than for our preparation. As with the studies on our rat adrenal cells, the ACTH threshold dose for ZF cells (canine whole adrenal cells) is lower than for ZG cells. A difference with canine cells is that the steroidogenic response shows a maximum at a certain concentration of angiotensin II (about 5×10^{-10} M) with a subsequent decrease above that concentration. This decrease, after a maximum response, was not found with ACTH. The canine adrenal cells have the advantage that cortisol production acts as a unique marker for zona fasciculata activity and it was stated to be negligible in Catt's preparation.

The rat adrenal cell and tissue responses reported show some differences with respect to K^+ concentration required for maximum output. Most reports give the maximum response at 8 mM $K^{+,58, 91, 116, 192}$ with a subsequent marked decrease at higher concentrations of the ion. However, although Catt and co-workers[112] have not published detailed results on their rat adrenal cells at different concentrations of K^+, they used 15 mM K^+ in some experiments with both canine and rat adrenal cells to study stimulation of steroidogenesis and cyclic AMP output (Section III.B.4) and obtained good steroidogenic stimulation at that concentration. However, Marusic and co-workers[194] found a maximum production for canine slices at 12 mM K^+ but the more usual 9 mM K^+ maximum for rat adrenal tissue. Braley & Williams[195] found a maximum for corticosterone output at about 8 mM K^+ but at 18 mM K^+ for aldosterone. In their later studies, as in those of Catt and co-workers, they included essential amino acids in the incubation medium. There is no explanation at present for these differences in K^+ concentration needed for maximum aldosterone stimulation, but it is clearly a characteristic of a particular cell preparation which must be taken into account when comparing results from different groups.

The steroidogenic potency of [des-Asp^1] angiotensin II (angiotensin III) is of considerable interest as it has been suggested that this compound rather than [Asp¹] angiotensin II is the real physiological messenger in the translation of the effects of salt depletion to aldosterone production,[196] although its peripheral plasma concentration is greater than that of [Asp¹] angiotensin II only in the rat.[197] From the earliest report from studies on the sheep adrenal transplant,[198] *in vivo* work indicated that angiotensin III was equipotent with angiotensin II in stimulating aldosterone. Certain *in vitro* results, including one by Goodfriend & Peach,[199] with rat adrenal cells supported this result. However, using rat adrenal cells, we found that although the maximal steroidogenesis achieved was equivalent for the same two peptides, angiotensin III was only about 10% as potent on an equimolar basis. Other groups,[200, 201] using rat cells and 2 h incubations subsequently reported similar results.[39] Catt and co-workers[200] have explained these results (except those of Peach[199]) by the finding that, surprisingly, the rat adrenal capsular preparation metabolizes angiotensin-like peptides at the C-terminal quite rapidly and the rate of catabolism is faster for angiotensin III than for angiotensin II. Therefore if short incubation times are used for the rat preparations, the steroidogenic activities of the two peptides become more equivalent. The potency of the two peptides is equal for a greater range of incubation times for canine adrenal cells when the metabolism of the peptides is less. This explanation seems reasonable but it implies that conditions *in vivo* correspond to those of lowered metabolism which is not necessarily so.[202] More quantitative data on the metabolism of the peptides both *in vivo* and *in vitro* are required. Various analogues of angiotensin II in addition to [des-Asp¹]-angiotensin II are active *in vitro*.[201,203,204] These include [Sar¹] angiotensin II which, except for the results of Peach & Chiu,[205] is more active than [Asp¹] angiotensin II.[201,203,204] As this N-terminal substitute is not metabolized by angiotensinase A, this indicates that the formation of [des-Asp¹] angiotensin II is not obligatory for biological activity.[204] Other angiotensin analogues are antagonists[206] with varying amounts of agonist activity.

The maximum response ratios for corticosterone (stimulated/basal outputs) for ACTH- and Hypertensin-stimulated rat ZF cells are about 120-fold compared with about 2-fold for K^+, ACTH, serotonin and Hypertensin for ZG stimulated cells. Aldosterone response is nearly 6-fold for these stimuli of ZG cells.[116] This illustrates a characteristic difference between zona glomerulosa and fasciculata tissue both *in vitro* and *in vivo* and corresponds to the response required for physiological function. Another difference is that whereas, as previously described, the zona fasciculata is very specific in response, effective stimuli being restricted to ACTH and closely related peptides, the ZG cells, like adipose cells,[207] are activated by a variety of stimuli. Ammonium, rubidium and caesium ions are active[58] presumably by a similar mechanism to that of K^+ although the maximum responses to these ions have not been compared. Compounds related to serotonin such as bufotenine, 5-methoxytryptamine and, at higher concentrations, L-5-hydroxytryptophan, effectively stimulate aldosterone production and this is inhibited by serotonin antagonists such as methysergide and a compound BC-105 which is related to cyproheptadine.[29, 31] The early failure to demonstrate significant aldosterone stimulatory activity of serotonin *in vivo* has usually been explained by its adsorption or uptake by platelets[29] but Edwards and co-workers[30, 31] have recently reported positive effects on aldosterone secretion in humans and rat. It is of interest that in studies of Edwards *et al.*,[78] bromocriptine (a dopamine agonist but possible serotonin antagonist) inhibited the action of both serotonin and angiotensin, suggesting that serotonin plays a role in the action of the peptide. However, serotonin but not angiotensin II stimulates cyclic AMP output (Section III.B.4) in most adrenal preparations.[112] Aldosterone secretion in the carcinoid syndrome, where the general biological activity of serotonin is evident, could be of interest in this regard. Recently, metoclopramide, a dopamine antagonist which was thought to act through increasing levels of prolactin *in vivo*, was shown to act in hypophysectomized human subjects[208] and directly on ZG cells. This was shown in a superfused system when dopamine also inhibited the stimulation by metoclopramide.[94] Dopamine and the dopaminergic agonist, bromocriptine, were shown to inhibit ACTH- and angiotensin II-stimulated aldosterone output in bovine adrenal cells.[209, 210] However, Brown *et al.*[33] did not obtain a direct effect of metoclopramide *in vitro* on bovine ZG cells although they reported an effect using minces of human adrenal adenomas and hyperplastic adrenals[211] and in this respect, the field is still somewhat confusing.

Flack *et al.*[188] using the *in vitro* superfusion system devised by us[17] were the first workers to show an effect of prostaglandins (E_1, E_2 and $F_{1\alpha}$) on adrenal glands, from hypophysectomized rats. Saruta & Kaplan[212] found an effect of prostaglandins (E_1 and E_2) on aldosterone synthesis in outer slices (zona glomerulosa) of beef adrenal tissue. Some prostaglandins (A, $F_{1\alpha}$ and $F_{2\alpha}$) were ineffective. Using the Florey sheep adrenal transplant,[213] the E-type was actually found to decrease aldosterone secretion. Spat and co-workers reported that prostaglandins of the E-type stimulated steroidogenesis by rat ZG tissue. These prostaglandins are rapidly inactivated in the pulmonary circulation and probably do not generally circulate[214] but recently the Hungarian group[71] have found that the A-type prostaglandins also stimulate aldosterone production by rat adrenal capsules. Unlike the E-type prostaglandins, however, the A-type stimulate the adrenal glands of salt deplete but not replete rats. Spat *et al.*[215] also studied the effect of prostaglandin synthetase inhibitors. These increased ACTH-stimulated corticosterone production rates in the capsule but reduced the corresponding stimulation rate in the decapsulated gland. The same authors[216] reported that the aldosterone increase after natriuresis (2 h after frusemide administration) is not obtained after pretreatment with indomethacin, a result not found in man by another group.[217] However, Campbell et al.[70] have reported that, in addition to their effect on renin release (Section I.D), prostaglandins also play a role in the angiotensin-mediated increase of aldosterone by ZG cells. Thus angiotensin II and III infusion in vivo increased prostaglandin release from adrenal capsules but the increase in aldosterone secretion in response to the angiotensins was reduced by the prostaglandin-synthesis inhibitors, indomethacin and meclofenamate. Additional evidence that this was a direct action on the adrenal gland was provided from in vitro incubations of capsular cells which showed that the addition of indomethacin not only reduced prostaglandin release but also inhibited the steroidogenic effects of angiotensin II and III. Meclofenamate had no effect on aldosterone stimulation by angiotensin II but markedly decreased the aldosterone response to angiotensin III. The reason for this differential action of indomethacin and meclofenamate was unexplained in these studies. These data are typical of those obtained in studies on the direct role of the prostaglandins on aldosterone secretion which seem to be variable, nonspecific and incapable of any general, logical analysis.

At the recent VIth International Congress of Endocrinology in Melbourne there were a number of reports on effects of substances and/or their antagonists or agonists on ZG cells. In addition to the effects of serotonin and dopamine already discussed, those of epinephrine, norepinephrine[218] and histamine[219] were reported. Although it must now be appreciated that the ZG cell response is relatively nonspecific, the number of substances suggested to be involved in its control is rather alarming and full publications and confirmatory work of these studies are awaited with interest.

Steroidogenesis by ZG cells is stimulated by a variety of factors; ACTH, angiotensin II and III (and their analogues) and serotonin (and related compounds), all stimulators being in the physiological range. The use of dispersed adrenal cells rather than tissue has, therefore, allowed the in vitro *demonstration of the stimulation of aldosterone by physiological concentrations of angiotensin II (Section I.C). Antagonists to angiotensin and serotonin decrease aldosterone stimulated production. K^+ (and other ions such as NH_4, Rb or Cs), also stimulate this hormone. Most workers report a maximum stimulation with 8 mM K^+ but some studies show a maximum at higher ionic concentrations. There are some reports of direct ZG cell steroidogenic stimulation by the dopamine antagonist, metoclopramide and inhibition by dopamine (and the dopaminergic agonist, bromocriptine). The direct effect of prostaglandins on ZG (and ZF) cell steroidogenesis is variable.*

(2) Effects on pathways of biosynthesis of aldosterone

Studies on dispersed adrenal cells have confirmed the work with adrenal tissue that an early biosynthetic pathway before pregnenolone is stimulated by all known factors which increase aldosterone production (see Section 1.D). Corticosterone output is increased together with aldosterone (Fig. 5) and after addition of cyanoketone derivatives which inhibit the conversion of pregnenolone to progesterone, two groups,[220, 221, 222] have found increases in pregnenolone production with angiotensin II, ACTH and K^+. It seems likely that, as with ZF cells and ACTH, the pathway affected is from cholesterol to pregnenolone and the most important mechanism appears to be on the interaction of cholesterol with cytochrome P-450 complex.[63]

Effects on a later pathway such as from corticosterone to aldosterone are more difficult to demonstrate. However, we[89, 223] found that stimulation of this pathway could be readily shown if albumin was used in the incubation medium. The aldosterone output of the cells in various stimulatory situations is not only correlated with increases in corticosterone output (Fig. 5), as would be expected if corticosterone were a biosynthetic intermediate, but the aldosterone/corticosterone ratio is proportional to the output of corticosterone.[116] It therefore seems that corticosterone activates its own conversion. This effect, which occurs rapidly and is not inhibited by cycloheximide,[223] automatically couples effects on the early and late pathways and may be important physiologically. It does, however, obscure more specific increases in the conversion of corticosterone to aldosterone. Nevertheless, it was considered that increases in the ratio aldosterone/corticosterone[2] would reveal more specific effects. K^+ concentrations in the physiological range of 5·6 and 5·9 mM increased the value significantly and also 13 mM but at 8·4 mM, angiotensin and ACTH did not increase the ratio and the effect of serotonin was only marginally significant statistically.[116]

A more direct approach has been adopted by the two groups previously mentioned in connection with work on the early pathway. Aguilera & Catt[222] found that the cyanoketone inhibitor did not inhibit conversion of corticosterone to aldosterone and therefore could be used to prevent endogenous production of corticosterone in studies on the late pathway. Addition of a suitable concentration of corticosterone then enables its concentration in the medium to be maintained constant. The production of aldosterone, therefore, indicates the conversion of corticosterone to aldosterone which was increased by 15 mM K^+, angiotensin II and ACTH in rat cells. In dog ZG cells, angiotensin II and K^+ but not ACTH stimulated the late pathway. Aguilera & Catt[222] found that aminoglutethimide, which inhibits the conversion of cholesterol to pregnenolone, also inhibited the conversion of corticosterone to aldosterone in both species. However, the conversion of deoxycorticosterone to aldosterone was inhibited only in the dog, possibly indicating an alternative late pathway in the rat not proceeding through corticosterone. McKenna et al.,[220, 221] surprisingly in view of the results of Catt et al., used aminoglutethimide to prevent the endogenous production of

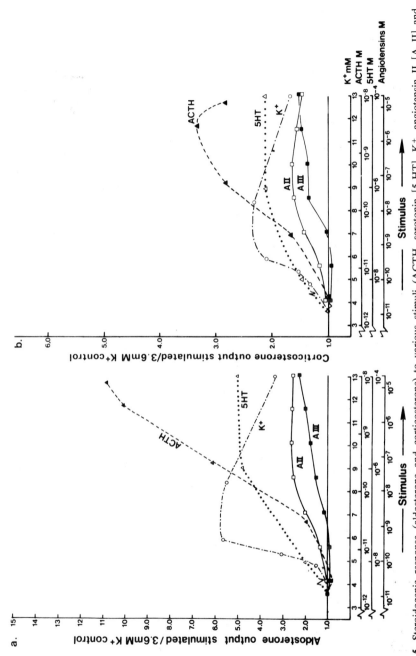

Fig. 5. Steroidogenic response (aldosterone and corticosterone) to various stimuli (ACTH, serotonin [5-HT], K⁺, angiotensin II [A II] and angiotensin III [des-Asp¹|angiotensin II [A III]) by rat capsular cells (95% ZG, 5% ZF cells).[116]

corticosterone. After addition of exogenous corticosterone and this inhibitor, McKenna *et al.*[220, 221] found that K^+ and angiotensin II increased the conversion of corticosterone to aldosterone.

They also found that from 0 to 6 mM K^+ the conversion of corticosterone is increased but is decreased from 6 to 12 mM. The early pathway is stimulated in both ranges of increased K^+ concentration,[221] which agrees with our results for cells of another species.[116] Neither the results of Catt *et al.* nor McKenna, showing a positive effect of angiotensin II on the late pathway, agree with ours[116] using the aldosterone/corticosterone[2] index, but as previously explained, only marked effects with the latter approach are probably reliable. From Muller's studies in the rat,[39] using *in vitro* incubations following *in vivo* stimulation, the explanation of these results might be that K^+ has a more marked action on the late pathway compared with the weaker effects of factors such as angiotensin II and ACTH.

All stimuli of aldosterone production seem to act on an early pathway before pregnenolone. The effects of K^+ concentration in the physiological range on the late pathway, corticosterone to aldosterone (and deoxycorticosterone to aldosterone), can now be readily demonstrated in vitro. There are probably similar effects (although rather weaker) with angiotensin II and possibly in some species with ACTH. If these specific stimuli and also alterations in sodium and potassium intake are applied in vivo and the biosynthetic pathways then are subsequently investigated in vitro, either in whole cells or in subcellular particles, effects on the late pathway can be demonstrated much more readily, although certain negative results have been recorded. However, the in vivo effects are usually established over a much longer period of time and it is always possible that the in vivo stimuli have direct natriuretic effects which may activate unknown biosynthetic pathways. It will be of interest to observe the effects of the late pathways of various agents not only with the stimuli discussed here, but with alterations in possible primary factors such as intracellular K^+ and Ca^{2+} and the effects of antagonists to their movements, such as lanthanum and ouabain, which will be mentioned later (Section III.B.5 & 6).

(3) Receptors

Studies with labelled angiotensin II have shown that high affinity receptors for the peptide are in the ZG cells.[224, 225, 226] Agonist and antagonist activities of angiotensin analogues correlate well with the relevant association constants[227, 228] and with biological activity. The topic has been critically reviewed by Regoli.[229] Catt and co-workers[230] found that changes in angiotensin II concentrations *in vivo* brought about by a variety of procedures, including changes in sodium and potassium intake, administration of angiotensin II[231] and lowering of angiotensin II by blockade of the enzymic system responsible for the conversion of angiotensin I,[232] lead to alterations of angiotensin II receptor characteristics. In the physiological range of concentrations of angiotensin II, the affinity of the receptors was changed in short term experiments and the number of sites at longer time intervals with the return of the affinity to normal values.[132] The overall effect is of increased binding of the peptide to adrenal cells with raised concentrations of angiotensin II tiself, and decreased binding at lowered concentrations of peptide, an example of the rare phenomenon of up-regulation, leading to positive feedback control. However, the number of sites could be altered chronically by increases in the size and number of zona glomerulosa cells. Changes in affinity of the sites are not usually involved in down-regulation. Meyer and co-workers[233, 234] found that administration of angiotensin II to nephrectomized rats decreased the binding of angiotensin II to uterine and adrenal receptors. Catt and co-workers[231] are inclined to ascribe this to the very large doses of peptide used as they have found that administration of such amounts lowers the number of receptor sites. However, Meyer and co-workers[234] also found that nephrectomy increased the adrenal binding of angiotensin II. This group measured the binding to a particulate fraction from the whole adrenal, whereas Catt and co-workers[231, 132] used ZG cells prepared from rat adrenal capsules and this may contribute to the discrepant results. Nevertheless, according to recent reports,[235, 236] increased K^+ concentration acted directly on the adrenal to raise the number (and equilibrium association constant) of angiotensin II receptors and this may be a factor in the apparent down-regulation observed after nephrectomy as K^+ concentration can then be increased.

Catt and co-workers[132] believe that the up-regulation of angiotensin II binding in salt restriction could explain the increase in sensitivity of the zona glomerulosa to angiotensin II in this situation. However, as previously discussed in Section I.C, Oelkers et al.[40] found that administration of angiotensin II, although increasing the sensitivity of human adrenal cells in the sodium replete state, cannot account for the total effect resulting after salt depletion. A possible explanation is that sodium restriction results in milder sensitization than salt depletion employing such procedures as diuresis.[39] This problem has also been discussed by Coghlan and co-workers[237] but they point out that the increased zona glomerulosa sensitivity in mild salt depletion can occur without any change in angiotensin II levels. Angiotensin II is then performing only a permissive role and cannot be a tropic adrenal factor such as in controlling receptor numbers or affinity.

In contrast to the effect on the adrenal receptors, the binding of angiotensin II to smooth muscle receptors is decreased in salt depletion (and with K^+ loading[235, 236]) and this corresponds to the lowering of its pressor activity.[131, 132]

However, the two types of receptors are similar according to Capponi & Catt.[225]

It is difficult to judge quantitatively from binding studies whether the increased sensitivity of the adrenal in vivo in salt restriction can be explained solely by alterations in the characteristics of the adrenal receptors with their up-regulation on changes in angiotensin II concentration. However, it now seems certain that part of this phenomenon of increased adrenal sensitivity can be explained in this manner. In contrast, the sensitivity of smooth muscle to the contractile effects of angiotensin II is decreased in salt depletion and there are corresponding changes in the binding of the peptide to the appropriate receptors.

(4) Cyclic AMP

Studies with ZG cells on the correlation of cyclic AMP and steroid values for any one stimulus have not been as extensive as those on ZF cells for ACTH. As the control of the zona glomerulosa is multiple, investigations on these cells are in a preliminary stage. Nevertheless, interesting if often discordant data are emerging.

In early studies[75, 238] we showed that with rat capsular adrenal cells, ACTH, K^+, serotonin and angiotensin II (Hypertensin at high concentrations) could all increase cyclic AMP. The nucleotide could also increase steroidogenesis when added to the incubation medium of unpurified and purified ZG cells.[75] However, as with other endocrine cells (Section III.A.3), in the case of all these agents steroidogenesis could be increased at low levels of stimulation without a significant rise in cyclic AMP output. With K^+ stimulation[239] the dichotomy was apparent at 5·9 mm with no significant increase in cyclic AMP but a near maximal increase in corticosterone output and at 8·4 mm when cyclic AMP rose sharply with only a slight further rise in corticosterone. At 13 mm there was the decrease in steroid output previously described (Section III.B.1), but a further small increase in nucleotide concentration from that found at 8·4 mm K^+. A kinetic study showed that 8·4 mm K^+ increased cyclic AMP markedly after 40 min but 5·9 mm K^+ did not raise the nucleotide in any time interval from the start of incubation.[239]

In contrast, Catt and co-workers[112] report no effect of K^+ on cyclic AMP output, with and without addition of phosphodiesterase inhibitor (MIX) to the incubation medium, in either rat or dog cells. Unfortunately aldosterone output could not be measured when MIX was added due to nonspecific effects on steroidogenesis. Presumably different concentrations of K^+ were used for the rat cells with similar negative results for cyclic AMP stimulation (although only data using 15 mm were published). There is no obvious explanation for the discrepancy.

Even more confusing are the results of various groups with the effects of angiotensin II on nucleotide output. We found that Hypertensin at high concentrations $(4 \times 10^{-4} \, \text{M})$ was effective in increasing cyclic AMP output.[75, 238] Also it was still effective, although the response was lower, after purification of the ZG cells. Later work showed that concentrations up to $4 \times 10^{-4} \, \text{M}$ of pure [Asn^1] or [Asp^1] angiotensin II, although increasing steroidogenesis, did not increase cyclic AMP output. Therefore it seems likely that the ACTH-like impurity in the preparation of Hypertensin which stimulates steroidogenesis in the zona fasciculata-reticularis was responsible for the increases in cyclic AMP in the earlier studies and that this contaminant might also stimulate nucleotide output in pure ZG cells (see the analogous discussion in Section III.A.1). Catt and co-workers[112] also report that pure [Asp^1] angiotensin II does not increase nucleotide output at any peptide concentration in dog and rat capsular adrenal cells either with or without the addition of MIX to the incubation medium. Extracellular, intracellular and receptor bound nucleotide were all not increased by angiotensin II or [des-Asp^1] angiotensin II in these studies. The Catt group[240] also reported that these stimuli, i.e. angiotensin II, [des Asp^1] angiotensin II, 15 mM K$^+$ and even ACTH had no effect on cyclic GMP output by dog zona glomerulosa cells. A similar lack of effect of angiotensin II on cyclic GMP output by rat ZG cells has been observed by us (unpublished observations).

Shima et al.,[161] using rat capsular adrenal tissue, obtained good response of aldosterone with a dose of Hypertensin $(10^{-5} \, \text{M})$ which probably did not stimulate the zona fasciculata contamination. Again this dose of angiotensin II, which stimulated steroidogenesis, did not increase cyclic AMP or cyclic GMP in the tissue. Phosphodiesterase was increased by angiotensin but the response of steroid and cyclic AMP to angiotensin II was not affected by the addition of theophylline. Peytremann[191] and Kaplan and co-workers[241] using bovine adrenal cells and tissue also found no increase in cyclic AMP with good stimulation of aldosterone with angiotensin II.

In contrast to our results and those of others,[112, 161, 241] Bing & Schulster[242] found a good correlation between cyclic AMP and steroidogenesis after stimulation with [Asp^1] angiotensin II in their rat capsular adrenal cell preparations. The reason for these differences is not clear. We (unpublished work) have examined different albumin concentrations in the medium (0·5% or 4%), times of incubation (60 min or 120 min) and assay systems (adrenal protein binding or immunoassay of acetylated nucleotide[149]) but still find no increase in cyclic AMP with [Asp^1] angiotensin II stimulation. Unfortunately, Bing & Schulster have not reported effects of other stimuli on cyclic AMP output which could be used as negative control experiments.

According to Valloton et al.[243] pure angiotensin II at $10^{-7} \, \text{M}$ stimulated cortisol production 5-fold in bovine ZF cells with again no increase in cyclic

AMP, contrary to an earlier report by Peytremann *et al.* who then also showed a positive cyclic AMP response.[117] Fujita *et al.*[112] found stimulation of steroidogenesis (but only a 50% increment from basal) with pure angiotensin II in canine ZF cells, but no increased cyclic AMP. With comparable stimulation of steroidogenesis, ACTH did increase cyclic AMP.[112]

We[238] found that different concentrations of ACTH increased nucleotide output in capsular rat adrenal preparations. Fujita *et al.*[112] observed this effect of ACTH (10^{-10} M to 10^{-7} M) on dog and rat adrenal cells and for extracellular and receptor bound nucleotide. It was possible that the interpretation of the early experiments of the Bethesda group with ACTH could have been affected by the presence of contaminating ZF cells, but we used purified ZG cells and still obtained positive nucleotide stimulation with ACTH[75, 238] and the dog cells used by Fujita *et al.* are presumably reasonably pure. Recently the Bethesda group[112] employed the column filtration method of McDougall & Williams[110, 111] to obtain pure ZG cells and still obtained cyclic AMP stimulation with ACTH.

In our studies,[238, 239] serotonin at different concentrations (from 10^{-7} M) stimulated nucleotide output but there was one dose (10^{-8} M) which caused nearly maximum steroidogenic response but no significant increase in nucleotide. However, studies on ZF and Leydig cells, previously discussed in Section III.A.3, have reduced the significance of such lack of correlation. The Bethesda group[112] also found an increase in nucleotide with serotonin in rat adrenal cells.

Fujita *et al.*[112] found that choleragen (although slow in action on cyclic AMP) as well as ACTH and serotonin could increase steroidogenesis and cyclic AMP outputs of dog and rat adrenal ZG cells and 8-bromo cyclic AMP added to the incubation medium increased steroidogenesis. These studies with dog adrenal cells, which contain minimal ZF cell contamination, also showed that the effects of K^+ or angiotensin II were additive to those of ACTH but not to one another, suggesting, according to these workers, as with the results of cyclic AMP outputs, that the mechanisms for the K^+ and angiotensin II stimulation of steroidogenesis are different from that of ACTH. Actually detailed examination of the paper of Catt and co-workers[112] shows that, without added cyclic AMP, the maximum aldosterone response is greater for 15 mM K^+ than for angiotensin II with ACTH giving the greatest response of all three stimuli. However, on addition of cyclic AMP (either exogenously or endogenously from choleragen) the response to K^+ is lower than for angiotensin and is even doubtfully significant after choleragen nucleotide stimulation. These results might indicate that K^+ does have some cyclic AMP stimulatory effects in their experiments, but a lower proportion of an alternative stimulation mechanism than for angiotensin II. The reason for the lack of detection of this cyclic AMP effect in direct experiments by Fujita *et al.*[112] may be due to the difficulty of using MIX in this system because it inhibits steroidogenesis nonspecifically. This might also affect the nucleotide response. The use of MIX (see Section III.A.3) has been probably the most important factor in work on the ZF cell in revealing correlations of nucleotide and steroidogenesis. However, in that system, a dose can be chosen which inhibits phosphodiesterase without nonspecifically affecting steroidogenesis. On the other hand, Fujita *et al.*[112] do obtain enhanced output of cyclic AMP with ACTH for glomerulosa steroidogenesis with addition of MIX.

We[239] showed that alterations in K^+ concentration (2–13 mM K^+) changed the maximum steroidogenic response (at a particular K^+ concentration) to agents which increased levels of cyclic AMP, i.e. serotonin, Hypertensin and cyclic AMP itself. This conclusively demonstrated that changes in K^+ concentration *in vitro* particularly from 2 to 5·9 mM K^+, can modify the action of cyclic AMP and may be a factor in the effects of sodium depletion in the rat when the plasma concentration of K^+ changes, or in the effects of K^+ administration in several species. Fredlund *et al.*[244] also found analogous alterations in steroidogenesis with alterations in K^+ concentration (0–7·5 mM) on pure angiotensin II responses. As this peptide does not increase cyclic AMP in this preparation, these results demonstrate that K^+ concentration can also modify a non-cyclic AMP mechanism, although in the same studies it was reported that stimulated steroidogenic responses with ACTH, which also increases cyclic AMP levels, are also modified by K^+.

Catt and we agree that ACTH and serotonin increase cyclic AMP even in purified cells. Also, these and other groups report that pure angiotensin II does not increase cyclic AMP. However, Bing & Schulster obtained a good correlation of nucleotide and steroid levels with angiotensin II. We report a positive effect on cyclic AMP output with increasing K^+ concentrations but Catt and co-workers cannot confirm this with their preparation at 15 mM K^+. When there are increases in cyclic AMP, there may be dissociation between nucleotide and steroid levels but presumably the same explanation as for ZF cells may apply. The conflicting results between groups would seem to be due to the particular properties of the preparations used either because of the preincubation or incubation conditions or to the state of the cells, including their parent animal species or strain of origin, e.g. the ZG cells used by Bing & Schulster are of different appearance and size from those employed by ourselves.

(5) Role of K^+ and the effect of ouabain

Potassium is an obligatory requirement in the incubation medium to demonstrate stimulation of aldosterone production *in vitro*.[58] Results *in vivo* in man, particularly those of A. Muller and co-workers in Geneva,[245], which showed changes in aldosterone secretion after administration of small amounts of K^+ without alterations in plasma K^+, suggested that intracellular K^+ might be important as a primary factor in the control of aldosterone. There was a period when the effects of changes in intracellular K^+ was suggested whenever the known factors such as angiotensin II, ACTH and K^+ concentration could not be correlated with aldosterone secretion.[246] The only direct evidence came from studies of Baumber et al.[246] who showed an increase in adrenal K^+ content after salt depletion and angiotensin II and KCl infusion. However, they analysed whole adrenals and as the volume and K^+ content of a ZF cell is about ten times that of a ZG cell, the data do not provide reliable evidence of the changes in the K^+ concentration of the smaller cell. Nevertheless, Boyd et al.[247] also reported that the K^+ content of the outer zone of rat adrenals containing ZG cells was increased on salt depletion. Mendelsohn, Mackie and co-workers,[248, 249] using first a capillary tube method and later one involving centrifuging dispersed cells through phthalate ester to eliminate extracellular fluid, found no changes in total intracellular K^+ content or concentration with serotonin on unpurified ZG cells and with serotonin and high concentrations of Hypertensin using purified ZG cells. High external K^+ (8·4 mM) did cause a small significant increase in intracellular K^+. However, there were similar effects on ZF cells for the increase in K^+ concentration with high levels of K^+ and this is to be expected with the inhibition of membrane ATPase by the ion. There was no correlation between the percentage increase in K^+ content or concentration and the rise in steroid output after stimulation with high external K^+.

Two groups[250, 251] used electron probe X-ray microanalysis to assay the K^+ and Na^+ content of the cells in rat adrenal slices. Szalay et al.[251] found that there was no change in the K^+ content of the zona glomerulosa and zona fasciculata and also the medulla after either Na^+-deficiency or a Na^+-rich diet. The Na^+ content of the zona glomerulosa and zona fasciculata was increased in the salt loaded rats but the medulla was unchanged. The other group[250] found that Na^+ was greater in the zona glomerulosa than the zona fasciculata but K^+ was distributed evenly. Again, the aldosterone production could not be correlated with the K^+ content in the zona glomerulosa.

The conclusion from all the more recent studies is that there are probably no specifically correlated changes in total intracellular K^+ with steroidogenesis in the zona glomerulosa. However, this does not eliminate the possibility of the importance of changes in ionic concentration in a hypothetical smaller compartment within the intracellular volume. At the present time, this cannot be investigated directly until there are improvements in electron and fluorescence probe techniques. For the present, indirect approaches such as measurement of ATPase and the use of ATPase inhibitors, which might be expected to alter all pools of intracellular K^+ concentration, have therefore to be used. The results of the use of ouabain *in vivo* with the sheep adrenal transplant[68] and with adrenal tissue[66, 67] have been described in Section I.D.

As was also evident with studies on adrenal tissue (Section I.D.), Mendelsohn & Mackie,[248] in our laboratory, using rat adrenal capsular cells found a biphasic steroidogenic response with increasing concentrations of ouabain with stimulation at low doses and inhibition at high doses of the glycoside. A maximum output of corticosterone above the control values occurred at 10^{-6} M. At 10^{-5} M there was no change in steroidogenesis and 10^{-4} M slightly but insignificantly inhibited, whereas 5×10^{-4} M and 10^{-3} M markedly inhibited steroidogenesis. Intracellular K^+, measured by the capillary technique, decreased continuously from 10^{-7} M. Therefore as regards basal output, there was no correlation between intracellular K^+ and steroid output, except at the highest concentrations of ouabain. This concentration $(5 \times 10^{-4}$ M), however, inhibited ZF cell steroidogenesis and therefore its effects were considered to be nonspecific. This dose inhibited completely the stimulation by serotonin and probably inhibited the effects of external high K^+ concentration. At 10^{-5} M the basal corticosterone output of both ZG and ZF cells was not affected, but the intracellular K^+ concentration was markedly depressed. This ouabain concentration did not markedly inhibit the stimulation of steroidogenesis by either serotonin or 8·4 mM K^+. The intracellular K^+ of the basal and serotonin-stimulated cells was decreased by this ouabain dose but not the raised value after external K^+ increase. Overall these results of Mendelsohn *et al.* do not support a role for alterations in any component of intracellular K^+ in controlling steroidogenesis.

Braley & Williams,[195] using the identical cell preparation, obtained similar results as regards the effect of different concentrations of ouabain on the basal production of corticosterone and aldosterone. Ouabain (10^{-4} M) was again the dose which did not cause either stimulation or inhibition of steroidogenesis. This dose almost completely inhibited the stimulation with angiotensin II (Hypertensin) and ACTH but did not inhibit ZF cell output. It should be noted that Mendelsohn *et al.*[248] used 5×10^{-4} M to achieve the same purpose but then obtained inhibition of ZF cell steroidogenesis. The inhibition of ZG cell steroidogenesis by K^+ stimulation obtained by Braley & Williams[195] was prevented by increasing the extracellular K^+ to 13 mM and the authors assume that the intracellular K^+ is then restored. The effect of varying extracellular K^+ on the ouabain-blocked angiotensin stimulation was not investigated. Braley & Williams ascribe the difference in results to the use of mixed cells by other groups and their measurement of corticosterone only. It seems more likely that the dose of ouabain required to inhibit the stimulated steroidogenic responses of ZG cells, but not that of the ZF cells is critical. The effect of low doses of ouabain on dog adrenal slices[252, 253] and homogenates[249] is only slightly stimulatory on aldosterone secretion unlike rat adrenals. Usually the action of the cardiac glycoside is to inhibit steroidogenesis. The stimulatory effect of angiotensin II on steroidogenesis is inhibited by doses of ouabain which presumably do not have nonspecific effects but both the Marusic[252] and Catt[240] groups report no effect of the peptide on Na^+–K^+ ATPase which presumably also indicates little effect of K^+ concentrations in that membrane compartment which involves ATPase. The field is in a confused state, but it does seem that, just as aldosterone does not function through a direct action on ATPase, so its control of production does not work specifically by such a mechanism.

Szalay[66, 254] observed that low concentrations of ouabain increased aldosterone output, presumably by inhibiting Na^+–K^+ ATPase and increasing extracellular/intracellular K^+ ratio. Therefore ethacrynic acid and chlorpromazine, inhibitors of Na^+–K^+ ATPase, should show increased aldosterone production. Actually 10^{-8} M ethacrynic acid increased aldosterone output but 10^{-3} M inhibited it. Chlorpromazine at any concentration failed to increase aldosterone secretion, which was explained by its Ca^{2+} antagonistic effect. Veratrine, a depolarizing agent, increased extracellular/intracellular K^+, and at 4×10^{-7} M, increased aldosterone secretion, whilst 4×10^{-5} M inhibited. This study illustrates vividly the problem with the use of inhibitors in such investigations without examination of nonspecific effects.

The measurement of membrane potential in ZG cells would clearly be of interest. However, the small size of the cells has presented technical obstacles to this aim. In the first of such studies published,[255] effects of K^+ and angiotensin II (in physiological concentrations) on resting membrane potential of kitten ZG cells in tissue slices were recorded. However, the potential as a function of concentration of these stimulators was very similar for both ZG and ZF cells. This suggests no specific mechanism for the generation of membrane potentials in ZG cells but does not exclude the existence of a specific biosynthetic apparatus in ZG cells responding to changes in membrane potential.

The overall conclusion of the results of these direct and indirect ionic studies indicates that, in total, intracellular K^+ is not correlated with steroidogenesis but this does not eliminate a role for a smaller pool of this ion. Another possibility, put forward by Braley & Williams, is that K^+ flux, not concentration, may be the relevant parameter. As they discussed, the biphasic nature of the ouabain stimulation of steroidogenesis in the ZG cell can be explained by a K^+ flux or ouabain-ATPase two-receptor theory. In the two-receptor theory, the ATPase activity and K^+ entry rate may be increased at low concentrations of ouabain. However, in ZG cells, the results of Mendelsohn show that intracellular K^+ is decreased at all concentrations of ouabain. Braley & Williams from their results also conclude that the K^+ flux is more likely to be applicable than the two-receptor hypothesis. However, a choice between K^+ flux or a hypothetical small pool being important cannot be *decided at the present time. Marusic and co-workers have indicated the importance of the concentration of K^+ on the external side of the membrane and that ouabain might block this effect by competing with K^+ active sites at the membrane. Membrane ATPase does not seem to be directly involved in the mechanism of stimulation. Membrane potential may be a key factor but the limited results obtained so far are not revealing as responses are common to both ZF and ZG cells.*

(6) Role of Ca^{2+}

As with K^+, the presence of Ca^{2+} in the extracellular medium is obligatory for basal production[58] and stimulation of steroidogenesis in ZG cells.[58, 256] The intracellular compared with the extracellular Ca^{2+} concentration is much lower and there is little possibility of its direct accurate estimation. Therefore at the present time, Ca^{2+} flux measurements are required. Of course, as with considerations of K^+ flux previously discussed (Section III.B.5) this might be, in any case, the important variable. Studies with Ca^{2+} transport inhibitors and ionophores may also be necessary.

Shima *et al.*[161] examined the role of Ca^{2+} in the control of steroidogenesis in the zona glomerulosa. They used rat adrenal capsular tissue, but their studies are so relevant to current studies on the corresponding dispersed cells, that a discussion of their results is included in this section. An elevation of Ca^{2+} concentration in the incubation medium was found to be effective in stimulating both aldosterone and cyclic AMP. However, theophylline produced a pronounced accumulation of cyclic AMP but no increase in steroid production. Angiotensin II (Hypertensin at 10^{-5} M), as previously discussed in Section III.B.4, increased steroidogenesis, but not cyclic AMP or adenylate cyclase levels in the tissue after 1 h incubations with and without theophylline, conditions which include those giving the maximum sensitivity for increase of cyclic AMP production in ZF cells (Section II.A.3). Increasing Ca^{2+} concentration raised the steroidogenic response to angiotensin II but the peptide did not further increase the raised cyclic AMP due to addition of extracellular Ca^{2+}. Tetracaine inhibited the basal aldosterone production and its response to angiotensin II at normal extracellular Ca^{2+} concentrations. This type of inhibitor blocks certain aspects of Ca^{2+} metabolism[177] but this has not been examined directly with exactly the same compound and at the same concentration as used in the adrenal tissue. Other Ca^{2+} antagonists, verapamil and lanthanum acted similarly on aldosterone production. These results suggest that intracellular Ca^{2+} but not cyclic AMP is important in controlling aldosterone production and its stimulation by angiotensin II in ZG cells. Unfortunately, these workers have not reported analogous results with other stimulators.[161]

Recently, Catt and co-workers,[256] using preparations of rat and dog adrenal ZG cells have studied in detail the requirements for Ca^{2+} of ACTH, angiotensin II and K^+ stimulated production of cyclic AMP and steroids. All three factors gave an increased output of aldosterone from 0 to 0·5 mM Ca^{2+} and with ACTH this continued at higher concentrations of the ion. Again, in these experiments only ACTH significantly increased cyclic AMP production. This was dependent on Ca^{2+} concentration and correlated with steroid output. Angiotensin II lowered and K^+ (14·5 mM) raised cyclic AMP output but neither effect was significant. Similar results were found after addition of MIX. When Ca^{2+} concentration was reduced, there was an increase in the ACTH concentration required for half-maximal steroid and cyclic AMP production and a decline in the maximal output of aldosterone. There was no change in the amount of stimulators required for half maximum steroid output but a decrease in the maximum with exogenous cyclic AMP, cholera toxin or serotonin. With angiotensin II and K^+ there was also no change in the amounts required for half maximum stimulation but the maximum steroid output was lowered. Binding of angiotensin II to ZG cell receptors was not altered in the absence of Ca^{2+} as has been concluded on more indirect evidence for the binding of ACTH to ZF cells. According to the authors, these findings also suggest that Ca^{2+} is required for the coupling of ACTH-receptor complexes with adenylate cyclase in the ZG cell. They also indicate an intracellular locus for Ca^{2+} subsequent to the action of cyclic AMP. The authors state that the Ca^{2+} requirement for ACTH action is quantitatively distinct from that for angiotensin II and K^+ consistent with a more Ca^{2+}-dependent mechanism of action for the latter regulators and these act subsequent to the initial coupling. However, it seems to us that without a prior knowledge of the lack of stimulation of cyclic AMP by angiotensin II and K^+,[112] the differences in their effects on Ca^{2+} metabolism compared with those of ACTH would not be impressive. More direct studies are required but the Bethesda group have reported these only in preliminary form.[162] As in the studies of Shima et al.[161] with angiotensin II, Fakunding[162] found in a preliminary report that lanthanum (1 μM) inhibited the stimulation of steroidogenesis in ZG cells by angiotensin II and K^+ but not by ACTH.

It is well known that inhibitors give results that are difficult to interpret. Lanthanum competes with Ca^{2+} for superficially located membrane binding sites and inhibits slow inward Ca^{2+} current in excitable cells, but effects on Na^+–Ca^{2+} transport systems are equivocal and species dependent.[257] Also, in intact cells, lanthanum does not penetrate intracellularly. Other Ca^{2+} antagonists may have nonspecific effects depending on the concentration used. Therefore, our group adopted a more direct approach, although this was usually much more difficult technically. In the first studies by Mackie et al.[164] using rat dispersed capsular adrenal cells, Ca^{2+} fluxes were studied by measuring $^{45}Ca^{2+}$

in the cells which had been centrifuged through phthalate ester oil to eliminate extracellular fluid. [^{3}H]Inulin was used to correct for the effect of any residual fluid. A significant effect on $^{45}Ca^{2+}$ uptake was observed when cells were exposed to a high K^+ concentration (8·4 mM) 30 min before addition of $^{45}Ca^{2+}$ but not if it was adjusted simultaneously. When the efflux of $^{45}Ca^{2+}$ from preloaded cells was examined by this static method, it was shown that the high K^+ medium inhibited faster than could be followed continuously. The content of $^{45}Ca^{2+}$ was consistently higher immediately and remained so by a constant margin. No such effect could be detected with ZF cells. These results are interesting and suggestive, but the significance of such rapid effects are not clear as the technique is so difficult and time consuming that adequate statistical data are almost impossible to collect as are adequate results to examine possible correlations of steroidogenic stimulating activity and effects on Ca^{2+} metabolism with different agents.

A more practical approach was to study the washout curve of preloaded $^{45}Ca^{2+}$ in the superfusion apparatus devised by McDougall & Williams and briefly described in Section III.A.4. The original apparatus of Lowry & McMartin,[76] which contained a mixture of Biogel and cells, was modified for $^{45}Ca^{2+}$ washout experiments to reduce the dead volume and eliminate plastics which adsorb Ca^{2+} (and also used Sephadex rather than Biogel.[110]). After about 1 h of superfusion and measuring $^{45}Ca^{2+}$ efflux continuously, steroidogenic substances including K^+ are added and the changes in Ca^{2+} efflux followed in repeated experiments. ACTH (3×10^{-8} M) and 8·4 and 5·6 mM K^+ (from control 3·6) had no effect on $^{45}Ca^{2+}$ efflux in ZG cells. Angiotensin II (pure [Asp^1] A II) at concentrations of 10^{-9} and 10^{-10} M had a significant effect on both steroidogenesis and Ca^{2+} efflux. A concentration of 10^{-11} M angiotensin II had no significant effect on either parameter. External cyclic AMP (4 mM) stimulated steroidogenesis to a greater extent than angiotensin II but had no effect on $^{45}Ca^{2+}$ efflux. In control experiments, $^{45}Ca^{2+}$ efflux of ZF cells did not alter with concentrations of ACTH, [Asp^1] angiotensin II and K which stimulated the steroid output of ZG cells although only ACTH stimulated steroidogenesis in the ZF cells.[77]

As in the studies of Shima and Catt and co-workers using Ca^{2+} transport inhibitors, effects on Ca^{2+} metabolism, as measured by $^{45}Ca^{2+}$ efflux, seem to be specifically involved with stimuli which, in a particular cell preparation, act by a non-cyclic AMP mechanism. However, it should be emphasized that at the present time this inverse correlation ($^{45}Ca^{2+}$ efflux and cyclic AMP stimulation, either with increased steroidogenesis) is being made with a restricted number of different cell preparations and stimuli. It is critical for the theory that the Bing & Schulster cell preparation which, unlike all others, shows a response of cyclic AMP to angiotensin II does not elicit nonpermissive changes in Ca^{2+} metabolism readily demonstrable by Ca^{2+} transport inhibitor and $^{45}Ca^{2+}$ washout techniques. This is not to suggest that there is not the usual permissive interaction between intracellular Ca^{2+} and cyclic AMP mechanisms in the zona glomerulosa as indicated by the studies of Catt and co-workers. These indicate that all stimulation, including the maximum steroid outputs, requires Ca^{2+} (as for ACTH in ZF cells) for coupling of the binding receptor and the adenylate cyclase complex and both cyclic AMP and non-cyclic AMP activators for subsequent responses, e.g. maximum steroid output after cyclic AMP increases with ACTH and serotonin. It remains to be seen whether the non-cyclic AMP mechanisms revealed by steroidogenic additivity experiments, Ca^{2+} inhibitor and $^{45}Ca^{2+}$ efflux techniques, are equivalent.

C. ZONA RETICULARIS (ZR) CELLS

(1) Rat adrenal cells

Unit gravity sedimentation of rat decapsulated adrenal cells has given reasonably pure (over 90% in terms of other steroidogenic cells) preparations of ZF and ZR cells.[13] There are no marker steroids produced for these cells as for aldosterone by ZG cells and cortisol, in most mammalian species, by ZF-ZR cells. However, it has been found that radioactive pregnenolone is preferentially converted to deoxycorticosterone rather than to total 11-oxygenated steroids (corticosterone plus 11-dehydrocorticosterone), by ZR compared with ZF cells. The radioactive deoxycorticosterone/(corticosterone plus 11-dehydrocorticosterone) ratio may therefore serve as a marker, although not absolute, to distinguish between the two types of cell. As the non-radioactive measurements for these steroids gave a similar pattern,[258] it seems that this ratio is characteristically higher for ZR than for ZF cells, and probably indicates lower 11β-hydroxylating activity for the ZR cells. 18-Hydroxydeoxycorticosterone was produced from radioactive pregnenolone in similar proportions to that of corticosterone plus 11-dehydrocorticosterone for the two types of cell and progesterone to that of deoxycorticosterone. Very small amounts of androstenedione were produced by both types of cell. 17α-Hydroxylated steroids are not produced in measurable amounts by the adrenal cells of this strain of rats (Sprague-Dawley). 18-Hydroxycorticosterone was not studied.[13, 258]

Although the amounts of androstenedione produced were small, the ratio androstenedione/(corticosterone plus 11-dehydrocorticosterone) was the only indication of the relative amounts of the androgen (or preandrogen) and glucocorticoid produced by the two types of cell. On this basis, the ZR cells preferentially produced the preandrogen, androstenedione, relative to that of the 11-oxygenated steroids. Using ACTH at a dose giving maximum steroidogenic stimulation, it was found that the stimulation of corticosterone was much greater for ZF than for ZR cells (130-fold compared with 20-fold[258]). With the same dose of ACTH, cyclic AMP output stimulation was also much greater for ZF than ZR cells (see Section III.A.3).[150]

The response of deoxycorticosterone to ACTH was also greater for ZF compared with ZR cells. However, for both ZF and ZR cells the response of corticosterone to ACTH was always higher than for deoxycorticosterone. The most obvious explanation of this finding was that 11β-hydroxylation was acutely increased to a major extent by ACTH, particularly as the conversion of labelled deoxycorticosterone to corticosterone was also increased.[258] This conclusion would be contrary to the theory of Stone & Hechter,[49] that ACTH acts mainly on the conversion of cholesterol to pregnenolone and only to a minor extent on 11β-hydroxylation. However, it was later found for both types of cell (but particularly for ZF cells) that ACTH, by increasing corticosterone production, decreased the conversion of radioactive corticosterone to 11-dehydrocorticosterone. This increased the ratio of corticosterone to deoxycorticosterone which could occur without necessarily any alteration in enzymic activity if the appropriate 11-dehydrogenase is an enzyme of limited and constant capacity. The overall effect of this phenomenon is to amplify the increase in production of steroid (reflected by deoxycorticosterone outputs) to give exaggerated corticosterone stimulation output ratios. As the effect is more evident in ZF compared with ZR cells, it will also amplify the difference in the stimulation ratios for corticosterone (but not for deoxycorticosterone) for the two types of cell. However, the ratio of (corticosterone plus 11-dehydrocorticosterone)/deoxycorticosterone is a more constant characteristic of a cell type, probably more accurately reflecting 11β-hydroxylating activity.[259]

Although certain LH and FSH preparations (supplied by N.I.H) stimulated steroidogenesis by the ZR cells and this might be relevant to morphological changes in these cells after gonadectomy,[260] these preparations of gonadotropin also stimulated steroidogenesis in ZF cells. However, they all gave some reaction with ACTH antibodies and it therefore seems that the effects may be due to ACTH contamination. HCG and human LH (prepared by P. Lowry), which contained undetectable amounts of ACTH on immunoassay, did not stimulate either ZF or ZR cells. Prolactin was also inactive in these acute experiments. Neither K^+ nor angiotensin II at physiological concentrations stimulated either type of cell.[258, 259] Therefore, although there may be chronic effects with humoral substances such as prolactin,[261, 262] at the present time no factor other than ACTH has been found which stimulates rat cells *in vitro* acutely.

(2) Guinea-pig adrenal cells

The major disadvantage in the study of rat adrenal cells, if one wishes to relate the results to preparations of human adrenal cells which are difficult to study directly, is that 17α-hydroxylation and the production of androstenedione (Section I.A) is low.

Preliminary studies[109] have been carried out using guinea-pig adrenals which might be expected not to have these disadvantages. Dispersed cells have been prepared using collagenase, after decapsulation, as for rat adrenals. These mixed ZF and ZR cells had good 17α-hydroxylating activity (they produced cortisol, 11-deoxycortisol and 17α-hydroxyprogesterone) and also made androstenedione in appreciable amounts, but very little dehydro-epiandrosterone. The ZF and ZR cells although in the same size range, could be separated using unit gravity sedimentation with a higher concentration of albumin than usual (see Section II.C.). The slowest and fastest moving cells were considerably enriched and the preparations of ZR cells contained about 10% ZF cell contamination. The purest preparations of both the ZF and ZR cells produced cortisol, corticosterone and androstenedione as judged by conversion from radioactive pregnenolone. It therefore seems likely that both types of cell and certainly the ZR cells produce androgens and glucocorticoids as proposed by Symington et al.[5] Also the appropriate ratios of pairs of (radioactive) steroids, deoxycorticosterone-corticosterone, deoxycortisol/cortisol and andro-stenedione/11β-hydroxyandrostenedione were higher for ZR compared with ZF cells, again suggesting lower 11β-hydroxylating activity by ZR cells. The (radioactive) androstenedione (preandrogen)/cortisol (glucocorticoid) ratio was significantly higher, in all three experiments performed, for ZR compared with ZF cells. In a further three similar experiments but measuring endogenous, non-radioactive steroids, preferential production of androstenedione was clearly observed, both before and after ACTH addition.

The results on the stimulation of individual steroids by ZF and ZR cells by ACTH were variable and accurate comparisons will require a purer preparation of cells. Preliminary studies of the Percoll separation show a lower steroidogenic response for ZR compared with ZF cells.

(3) General conclusions on ZR cells

Studies with the rat and guinea-pig cells show that: (a) ZR and ZF cells produce the same steroids qualitatively; (b) there is preferential production of deoxycorticosterone compared to total 11-oxygenated steroids in the ZR cells, indicating relatively low 11β-hydroxylation in these cells. Androstenedione, which is a preandrogen, is also produced preferentially relative to total 11-oxygenated steroids (glucocorticoids) by the ZR cells; (c) as shown rigorously in the rat and in preliminary studies with guinea-pig cells, ACTH-steroidogenic stimulation is lower for ZR than for ZF cells.

The data, at least when considered quantitatively, are in agreement with two seemingly opposing hypotheses; that of Symington, who proposed that the zona reticularis and fasciculata produced both glucocorticoids and androgens and that of Vines, who proposed that only the zona reticularis produces androgens. The present data suggest, in accordance with Symington's theory, that the ZF and ZR cells both produce the two types of steroid but nevertheless the ZR cells tend to make preandrogens preferentially. These data are not in accordance with the views of Chester Jones who suggested that the zona reticularis could be vegetative and senescent.

Other in vivo work, as previously discussed in Section I.C., suggests that dehydroepiandrosterone sulphate is produced more specifically by the zona reticularis than androstenedione. These results were obtained with human subjects as only the higher primates produce dehydroepiandrosterone and its sulphate in substantial amounts. If it is accepted that the preandrogens are produced preferentially by the zona reticularis, there is therefore other evidence to suggest that ACTH stimulates the zona reticularis to a lesser extent than the zona fasciculata as ACTH increases androstenedione (or dehydroepiandrosterone) production to a lesser extent than cortisol (or corticosterone). This is particularly evident in the in vivo results of Parker & Odell with the adrenal output of the castrated dog.

The problem of the preferential increase in preandrogens in the human adrenarche has not been greatly elucidated by our studies with dispersed cells, e.g. the pituitary extract (used by Parker & Odell) does not preferentially stimulate preandrogen or zona reticularis steroidogenesis in rat and guinea pig cells. An alternative theory put forward by Anderson that a different mode of stimulation, e.g. frequency of ACTH output, might alter the preandrogen/ cortisol ratio in the adrenarche is also not supported by the present data on cells. Although ACTH may give a higher maximum stimulation for glucocorticoid rather than preandrogen output, the threshold sensitivities for ACTH are similar for the two types of steroid in dispersed cells as are the quantities for half maximum stimulation. Therefore the ratio of the two types of steroid does not alter with dose of ACTH. Obviously, cells from human adrenals may respond differently but the present results do not support either the Odell or Anderson type of explanation. Another hypothesis could be that specific hypertrophy of the zona reticularis could increase the secretion of preandrogens relative to glucocorticoids as the ZR cells produce the former preferentially. Such hypertrophy could be brought about by the chronic influence of a factor such as prolactin whose activity would not be revealed by the acute experiments with whole cells described here.

IV. General Considerations

A. SPECIFICITY OF STEROID RESPONSE

The three main types of cell of the adrenal cortex, from the zonae fasciculata, reticularis, and glomerulosa produce, in common, certain steroids such as progesterone, deoxycorticosterone, 18-hydroxycorticosterone, 18-hydroxy-deoxycorticosterone and more importantly corticosterone. However, their response to various stimuli differs both qualitatively and quantitatively for the three types of cell (Fig. 6). This allows the study of the mechanism of action of different stimuli in the ZG, ZF and ZR cells using the same end point, usually corticosterone output. According to present knowledge, only the conversion of corticosterone to aldosterone in the zona glomerulosa is an independent additional locus of action and cannot be regarded as the same end point. It is however a minor factor quantitatively *in vitro*.

Specificity of response of a steroid hormone to the various stimuli is achieved by: (a) The production of unique products such as aldosterone by the zona glomerulosa. In certain species such as man and guinea-pig, cortisol is produced by the zona fasciculata, and probably also by the zona reticularis, but not by the zona glomerulosa. The preandrogen, androstenedione, is also produced by both the zona fasciculata and zona reticularis but it is not known for certain whether it is made by the zona glomerulosa although this is very unlikely; (b) Certain steroids are produced in common by all cell types, as just described, but their response characteristics differ probably due to the quantity and characteristics of their plasma membrane receptors; (c) An amplification mechanism which exaggerates these specificity effects due to receptor properties by coupling early and late pathways. This amplifies, at least in the rat, the differential production of steroid to give greater stimulation ratios for corticosterone in the zona fasciculata and to a lesser extent in the zona reticularis. A similar effect occurs in the zona glomerulosa when the corticosterone level controls its own conversion to aldosterone and therefore aldosterone output is approximately proportional to the square of the corticosterone outputs. The mechanisms of these amplification effects in the zona fasciculata-reticularis and zona glomerulosa may be similar as they have several characteristics in common. In the zona fasciculata-reticularis, amplification is achieved because of the conversion of a maximum constant amount of corticosterone to 11-dehydrocorticosterone. The detailed mechanism of the amplification effect in the zona glomerulosa has yet to be elucidated. Similar changes in 11β-hydroxy to 11-oxo conversions have not been detected. However, it should be remembered that specific binding of steroids by a protein of limited capacity present in the cell could give rise to similar effects by an analogous mechanism. (d) Specific effects on the late

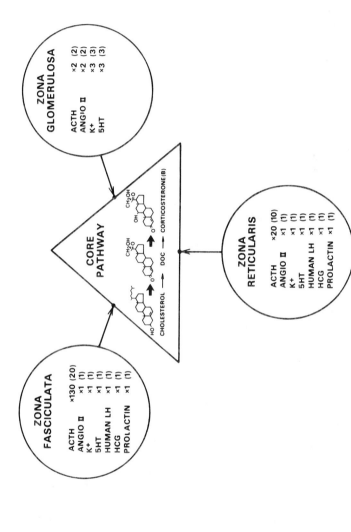

Fig. 6. Response of corticosterone (the steroid product of the core pathway) to various stimuli of the zonae fasciculata, reticularis and glomerulosa. Values without parentheses are stimulated output ratios (stimulated/basal outputs); values within parentheses are estimated stimulated production ratios (stimulated/basal production rates).

pathway, corticosterone to aldosterone. Such an effect both *in vitro* and *in vivo*, has been established for K^+, is probable for angiotensin II and possible for ACTH. No such specific effect on late pathways to corticosterone or cortisol (e.g. on 11β-hydroxylation) has been found in the zonae fasciculata and reticularis.

The overall conclusion from present results on the three types of adrenocortical cell is that the response of a steroid to different stimuli depends not on its chemical structure (although this may determine whether it is produced at all by a particular zone), but mainly on the cell type of origin.

B. MECHANISM OF STEROID STIMULATION

In most cells, cyclic AMP and Ca^{2+} control mechanisms are coupled. In systems which Berridge[163] and others have described as unidirectional, as the same control mechanism operates on or off, the changes in cyclic AMP usually control Ca^{2+} transport as shown by the effects on the $^{45}Ca^{2+}$ efflux of preloaded cells. These could be due to changes in the Ca^{2+} influx or efflux permeability properties of the plasma membrane or, more likely, to alterations in intracellular transport of the ion, but whatever mechanism operates it probably indicates alterations in available cytoplasmic Ca^{2+} concentrations.

An exception to this coupling of the two mechanisms in unidirectional systems seems to be the exocrine pancreas, where acetylcholine may release the secretion of enzymes such as amylase through increases in intracellular Ca^{2+} without changes in cyclic AMP.[163] However, in this complicated system with the possibilities of both release and biosynthesis of the product occurring and the presence of a mixture of cells, results can be ambiguous.[163] Nevertheless, recent results of Peterson[263] clearly indicate that, in the effects of agents such as acetylcholine, a Ca^{2+} mechanism is involved not only in the exocytosis of enzyme secretion but also in the direct ionic effects on membrane permeability. For many cells, effects of a high extracellular concentration of K^+ usually result in depolarization of cells, increased Ca^{2+} influx and effects which bypass the usual cyclic AMP mechanisms and also include no changes in nucleotide levels. However, these effects take place only with nonphysiological concentrations of K^+. In all other endocrine situations of physiological significance which have been studied, cyclic AMP takes part in control mechanisms and, at least to a major extent, acts by affecting Ca^{2+} transport. This is not to suggest that either the nucleotide or divalent ion cannot act independently but that the coupling between these mechanisms is usually also important. The relative importance of the two independent mechanisms and their interactions seems to depend on the type of cell.

In the ZF (and probably also the ZR) cell, the cyclic AMP mechanism is dominant although independent Ca^{2+} and certain nucleotide-Ca^{2+} interaction mechanisms may also be present. However, in the ZF cells, although ACTH according to recent results increases cyclic AMP levels in a correlated manner with steroidogenesis, it does not alter Ca^{2+} efflux as do most nucleotide stimulators in other cells. Normal concentrations of Ca^{2+} may be required for full stimulation but this may be a permissive effect with no changes in Ca^{2+} transport involved even though Ca^{2+} transport inhibitors show effects. The latter results may be difficult to interpret.

In the ZG cell, at least as regards one physiologically important controlling factor, angiotensin II, the situation is different from that found in most other cell systems. Stimulation with ACTH and serotonin is again dominated by the cyclic AMP mechanism and these alterations in nucleotide levels are not accompanied by effects on Ca^{2+} efflux. Also exogenous cyclic AMP stimulates steroidogenesis without affecting Ca^{2+} efflux. However, angiotensin II can increase steroidogenesis in the ZG cell without any effect on cyclic AMP and effects on steroidogenesis and on Ca^{2+} efflux are then correlated. Ca^{2+} transport inhibitors, such as lanthanum, in a similar preparation prevent this stimulation of steroid output by angiotensin II. In the same preparation, K^+ also does not stimulate cyclic AMP but increases steroid output and, again according to indirect inhibitor studies, affects Ca^{2+} transport. However, as in this study 15 mM K^+ was used, it remains to be seen whether this is the effect shown in many cells when such non-physiological concentrations of K^+ cause depolarization and subsequent Ca^{2+} uptake. In adrenal cells this could then stimulate steroidogenesis with the bypassing of the usual cyclic AMP mechanism. In another superfused ZG cell preparation, K^+ at physiological concentrations of 5·9 and 8·4 mM, which raised nucleotide output in static systems, increased steroid output but had no effect on Ca^{2+} efflux. Similar results were also observed with ACTH and serotonin.

Therefore, using physiological quantities of stimulating factors, in the unidirectional system of the ZG (and ZF) cell, the nonpermissive Ca^{2+} and nucleotide mechanisms, as indicated by the particular techniques employed, may be uncoupled whether the dominant mechanism is by alterations in Ca^{2+} or nucleotide levels. This occurs in a reasonably homogeneous cell population and following a simple activity of the cell, steroidogenesis.

In bidirectional systems, the stimulation and recovery stages may work by different mechanisms, e.g. in smooth muscle, contraction may be controlled by a Ca^{2+} mechanism without changes in cyclic AMP and recovery involve cyclic AMP as the dominant second messenger with the nucleotide affecting Ca^{2+} transport and restoring intracellular stores of the ion.[264] It is of interest that angiotensin II, which stimulates both smooth muscle contractions and zona glomerulosa steroidogenesis seems to be particularly potent in affecting Ca^{2+} transport not only in ZG cells but also in muscle[265, 266] and soft tissue generally[266] without necessarily altering cyclic AMP levels. Unfortunately, the mechanism of action of angiotensin II in smooth muscle cells has not been examined in the same detail as the present ZG cell investigations, although Freer[267] reports that verapamil, the Ca^{2+} transport inhibitor, inhibits angiotensin II-induced uterine smooth muscle contraction. The relevant definitive studies must await the availability of appropriate dispersed cells such as those prepared from uterine muscle tissue.

Also of interest, although not yet of proven physiological significance, is the effect of angiotensin II (and also catecholamines, through α adrenergic receptors, and vasopressin) on isolated hepatocytes. These hormones control certain aspects of carbohydrate metabolism in these cells through a Ca^{2+}-requiring cyclic AMP-independent pathway that leads to the phosphorylation of important regulatory enzymes.[157] Angiotensin II, which has a variety of extra-adrenal biological functions,[229] seems to act usually, although not exclusively, through such a mechanism which is being increasingly reported to operate in non-endocrine tissues by groups such as Garrison,[268] Cohen[269] Larner,[270] Hems,[271] and Robison & Sutherland.[272] Some of our recent work is reported elsewhere.[273]

ACKNOWLEDGEMENTS

Our colleagues in the Physics Department, particularly Drs J. Bradley, P. Hyatt, B. Williams and T. Powell contributed many of the experimental results (including unpublished work) and a significant part of the interpretations which however are our major responsibility. Dr P. Hyatt also helped to compose and execute some of the diagrams. We had the recent invaluable experimental technical assistance of Messrs D. Atkinson, K. Bhatt, F. Chu and A. Zananiri. Messrs D. Atkinson and S. Nightingale were responsible for the diagrams and photography. Mrs Marie Stuart and Miss Pamela Winter performed the typing and carried out some literature research. We would also like to thank Dr H. Rasmussen for some valuable comments. Our work has been supported by a Medical Research Council Programme Grant No. G969204C and a Royal Society Grant.

REFERENCES

1. Tait, J. F. & Tait, S. A. S. (1979). Recent perspectives on the history of the adrenal cortex. *J. Endocrinol.* **83**, 3P–24P.
2. Hayano, M., Saba, N., Dorfman, R. I. & Hechter, O. (1956). Some aspects of the biogenesis of adrenal steroid hormones. *Recent Prog. Horm. Res.* **12**, 79–123.
3. Ulick, S., Nicolis, G. N. & Vetter, K. K. (1964). Relationship of 18-hydroxycorticosterone to aldosterone. In *Aldosterone* (Baulieu, E. E. & Robel, P., eds.) pp. 3–17, Blackwell, Oxford.
4. Melby, J. C., Dale, S. L., Grekin, R. J., Gaunt, R. & Wilson, T. E. (1972). 18-hydroxy-11-deoxycorticosterone (18-OH-DOC) secretion in experimental and human hypertension. *Recent Prog. Horm. Res.* **28**, 287–351.

5. Symington, T. (1969). The adrenal cortex. In *Functional Pathology of the Human Adrenal Gland.*, pp. 3–216. Livingstone, Edinburgh & London.

6. Baird, D. T., Horton, R., Longcope, C. & Tait, J. F. (1968). Steroid prehormones. *Perspect. Biol. Med.* **11**, 384–421.

7. Weliky, I. & Engel, L. L. (1963). Metabolism of progesterone-4-C[14] and pregnenolone-7α-H[3] by human adrenal tissue. *J. Biol. Chem.* **238**, 1302–1307.

8. Dorfman, R. I. & Ungar, F. (1965). Biosynthesis of steroid hormones. In *Metabolism of Steroid Hormones*, pp. 126–128. Academic Press, New York and London.

9. Fattah, D. I., Whitehouse, B. J. & Vinson, G. P. (1977). Biosynthesis of aldosterone from 18-hydroxylated precursors by rat adrenal tissue *in vitro*. *J. Endocrinol.* **75**, 187–195.

10. Neher, R. (1979). Aldosterone: chemical aspects and related enzymology. *J. Endocrinol.* **81**, 25P–35P.

11. Luttrell, B., Hochberg, R. B., Dixon, W. R., McDonald, P. D. & Lieberman, S. (1972). Studies on the biosynthetic conversion of cholesterol into pregnenolone. *J. Biol. Chem.* **247**, 1462–1472.

12. Deane, H. W. (1962). The anatomy, chemistry and physiology of adrenocortical tissue. In *Handbuch der experimentellen pharmakologie* vol. 14, part 1, pp. 1–185. Springer-Verlag, Berlin, Gottingen and Heidelberg.

13. Bell, J. B. G., Gould, R. P., Hyatt, P. J., Tait, J. F. & Tait, S. A. S. (1978). Properties of rat adrenal zona reticularis cells: preparation by gravitational sedimentation. *J. Endocrinol.* **77**, 25–41.

14. Giroud, C. J. P., Stachenko, J. & Venning, E. H. (1956). Secretion of aldosterone by the zona glomerulosa of rat adrenal glands incubated *in vitro*. *Proc. Soc. Biol. Med.* **92**, 154–158.

15. Ayres, P. J., Gould, R. P., Simpson, S. A. & Tait, J. F. (1956). The *in vitro* demonstration of differential corticosteroid production within the ox adrenal gland. *Biochem. J.* **63**, 19P.

16. Ayres, P. J., Garrod, O., Tait, S. A. S. & Tait, J. F. (1958). Primary aldosteronism (Conn's syndrome). In *An International Symposium on Aldosterone* (Muller, A. F. & O'Connor, C. M., eds.), pp. 143–154. Churchill, London.

17. Tait, S. A. S., Schulster, D., Okamoto, M., Flood, C. & Tait, J. F. (1970). Production of steroids by *in vitro* superfusion of endocrine tissue. II Steroid output from bisected whole, capsular and decapsulated adrenals of normal intact, hypophysectomized and hypophysectomized-nephrectomized rats as a function of time of superfusion. *Endocrinology* **86**, 306–382.

18. Marusic, E. T. & Mulrow, P. J. (1969). 18-hydrocorticosterone biosynthesis in an aldosterone secreting tumor in the surrounding nontumorous adrenal gland. *Proc. Soc. Exp. Biol. Med.* **131**, 778–780.

19. Griffiths, K. & Glick, D. (1966). Determination of the 11β-hydroxylase activity in microgram samples of tissue: its quantitative histological distribution in the rat adrenal, and the influence of corticotrophin. *J. Endocrinol.* **35**, 1–12.

20. Vines, H. W. C. (1938). The adreno-genital syndrome: some histological observations. In *The Adrenal Cortex and Intersexuality* (Broster, L. R., Allen, C., Vines, H. W. C., Patterson, J., Greenwood, A. W., Marrian, G. F. & Butler, G. C., eds.), pp. 137–197. Chapman & Hall, London.

21. Jones, T. & Griffiths, K. (1968). Ultramicrochemical studies on the site of formation of dehydroepiandrosterone sulphate in the adrenal cortex of the guinea pig. *J. Endocrinol.* **42**, 559–565.

J. F. TAIT, S. A. S. TAIT AND J. B. G. BELL

22. Kime, D. E., Vinson, G. P., Major, P. W. & Kilpatrick, R. (1980). Adrenal-gonad relationships. In *General, Comparative and Clinical Endocrinology of the Adrenal Cortex* (Chester Jones, I. & Henderson, I. W., eds.), vol. 3. Academic Press, London (in press).
23. Ward, P. J. & Grant, J. K. (1963). The metabolism of [4-^{14}C]progesterone by adrenocortical tissue from patients with Cushing's syndrome. *J. Endocrinol.* **26**, 139–147.
24. McDougall, J. G., Coghlan, J. P., McGarry, E. E. & Scoggins, B. A. (1976). Aldosterone production by sheep adrenal glands *in vitro. J. Steroid Biochem.* **7**, 421–428.
25. Horton, R. & Tait, J. F. (1967). *In vivo* conversion of dehydroisoandrosterone to plasma androstenedione and testosterone in man. *J. Clin. Endocrinol. Metab.* **27**, 79–88.
26. Grumbach, M. M., Richards, G. E., Cato, F. A. & Kaplan, S. L. (1978). Clinical disorders of adrenal function and puberty: an assessment of the role of the adrenal cortex in normal and abnormal puberty in man and evidence for an ACTH-like pituitary adrenal androgen stimulating hormone. In *The Endocrine Function of the Human Adrenal Cortex* (James, V. H. T., Serio, M., Guisti, G. & Martini, L., eds.), pp. 583–612, Academic Press, London and New York.
27. Parker, L. N. & Odell, W. D. (1979). Evidence for existence of cortical androgen-stimulating hormone. *Am. J. Physiol.* **236**, E616–E620.
28. Komor, J. & Muller, J. (1979). Effects of prolonged infusions of potassium chloride, adrenocorticotrophin or angiotensin II upon serum aldosterone concentration and the conversion of corticosterone to aldosterone in rats. *Acta Endocrinol. (Copenhagen)* **90**, 680–691.
29. Muller, J. & Ziegler, W. H. (1968). Stimulation of aldosterone biosynthesis *in vitro* by serotonin. *Acta Endocrinol. (Copenhagen)* **59**, 23–35.
30. Mantero, F., Opocher, F., Armanini, D., Boscaro, M. & Edwards, C. R. W. (1979). Effect of serotonin (5-HT) on plasma aldosterone in man. *Acta Endocrinol. (Copenhagen) Suppl. 225,* **91**, 345.
31. Al-Dujaili, E. A. S., Boscaro, M., Espiner, E. A. & Edwards, C. R. W. (1980) *In vitro* and *in vivo* effects of indoleamines on aldosterone biosynthesis in the rat. *Abst. Int. Congr. Endocrinol.* VIth Melbourne, Australia No. 400.
32. Sen, S., Bravo, E. L. & Bumpus, F. M. (1977). Isolation of a hypertension-producing compound from normal human urine. *Circ. Res. Suppl. I* to **40**, 5–10.
33. Brown, R. D., Wisgerhof, M., Carpenter, P. C., Brown, G., Jiang, N.-S., Kao, P. & Hegstad, R. (1979). Adrenal sensitivity to angiotensin II and undiscovered aldosterone stimulating factors in hypertension. *J. Steroid Biochem.* **11**, 1043–1050.
34. Palmore, W. P. & Mulrow, P. J. (1967). Control of aldosterone secretion by the pituitary gland. *Science* **158**, 1482–1484.
35. Page, R. B., Boyd, J. E. & Mulrow, P. J. (1974). The effect of alpha-melanocyte stimulating hormone on aldosterone production in the rat. *Endocrine Res. Commun.* **1**, 53–62.
36. Vinson, G. P., Whitehouse, B. J., Dell, A., Etienne, T. & Morris, H. R. (1980). Characterisation of an adrenal zona glomerulosa-stimulating component of posterior pituitary extracts as *a*-MSH. *Nature (London)* **284**, 464–467.
37. Campbell, W. B., Brooks, S. N. & Pettinger, W. A. (1974) Angiotensin II- and angiotensin III-induced aldosterone release *in vivo* in the rat. *Science* **184**, 994–996.

38. Oelkers, W., Brown, J. J., Fraser, R., Lever, A. F., Morton, J. J. & Robertson, J. I. S. (1974). Sensitization of the adrenal cortex to angiotensin II in sodium-deplete man. *Circ. Res.* **34**, 69–77.
39. Coghlan, J. P., Blair-West, J. R., Denton, D. A., Fei, D. T., Fernley, R. T., Hardy, K. J., McDougall, J. G., Puy, R., Robinson, P. M., Scoggins, B. A. & Wright, R. D. (1979). Control of aldosterone secretion. *J. Endocrinol.* **81**, 55P–67P.
40. Oelkers, W., Schoneshofer, M., Schultze, G., Brown, J. J., Fraser, R., Morton, J. J., Lever, A. F. & Robertson, J. I. S. (1975). Effect of prolonged low-dose angiotensin II infusion on the sensitivity. of adrenal cortex in man. *Circ. Res. Suppl. I* to **36** and **37**, 49–56.
41. Grahame-Smith, D. G., Butcher, R. W., Ney, R. L. & Sutherland, E. W. (1967). Adenosine 3',5'-monophosphate as the intracellular mediator of the action of adrenocorticotropic hormone on the adrenal cortex. *J. Biol. Chem.* **242**, 5535–5541.
42. Haynes, R. C. & Berthet, L. (1957). Studies on the mechanism of action of the adrenocorticotropic hormone. *J. Biol. Chem.* **225**, 115–124.
43. Hechter, O. & Halkerston, I. D. K. (1964). VI. On the action of mammalian hormones. In *The Hormones* (Pincus, G., Thimann, K. V. & Astwood, E. B., eds.), vol. V, pp. 697–825. Academic Press, New York and London.
44. Harding, B. W. & Nelson, D. H. (1963). Dissociation of adrenal TPNH metabolism from ACTH stimulation. *Clin. Res.* **11**, 84.
45. Garren, L. D., Gill, G. N., Masui, H. & Walton, G. M. (1971). On the mechanism of action of ACTH. *Recent Prog. Horm. Res.* **27**, 433–478.
46. Rae, P. A., Gutmann, N. S., Tsao, J. & Schimmer, B. P. (1979). Mutations in cyclic AMP-dependent protein kinase and corticotropin (ACTH)-sensitive adenylate cyclase affect adrenal steroidogenesis. *Proc. Natl. Acad. Sci. U.S.A.* **76**, 1896–1900.
47. Schulster, D., Tait, S. A. S., Tait, J. F. & Mrotek, J. (1970). Production of steroids by *in vitro* superfusion of endocrine tissue. III. Corticosterone output from rat adrenals stimulated by ACTH or cyclic AMP and the inhibitory effect of cycloheximide. *Endocrinology* **86**, 487–502.
48. Catt, K. J. & Dufau, M. L. (1977). Peptide hormone receptors. *Annu. Rev. Physiol.* **39**, 529–557.
49. Stone, D. & Hechter, O. (1954). Studies on ACTH action in perfused bovine adrenals: the site of action of ACTH in corticosteroidogenesis. *Arch. Biochem. Biophys.* **51**, 457–469.
50. Karaboyas, G. C. & Koritz, S. B. (1965). Identity of the site of action of 3',5'-adenosine monophosphate and adrenocorticotropic hormone in corticosteroidogenesis in rat adrenal and beef adrenal cortex slices. *Biochemistry* **4**, 462–468.
51. Koritz, S. B. & Hall, P. F. (1964). End product inhibition of the conversion of cholesterol to pregnenolone in an adrenal extract. *Biochemistry* **3**, 1298–1304.
52. Simpson, E. R. & Mason, J. I. (1976). Molecular aspects of the biosynthesis of adrenal steroids. *Pharmacol. Ther. B* **2**, 339–369.
53. Pederson, R. C. & Brownie, A. C. (1980). Adrenocortical response to corticotrophin is potentiated by some part of the amino terminal region (16K fragment) of pro-corticotrophin/endorphin. *Proc. Natl. Acad. Sci. U.S.A.* (in the press).

54. Lefkowitz, R. J., Roth, J. & Pastan, I. (1970). Effect of calcium on ACTH stimulation on the adrenal: Separation of hormone binding from adenyl cyclase activation. *Nature* (*London*) **228**, 864–866.
55. Kelly, L. A. & Koritz, S. B. (1971). Bovine adrenal cortical adenyl cyclase and its stimulation by adrenocorticotropic hormone and NaF. *Biochim. Biophys. Acta* **237**, 141–155.
56. Grant, J. K., Symington, T. & Duguid, W. P. (1957). Effect of adrenocorticotropic therapy on the *in vitro* 11β-hydroxylation of deoxycorticosterone by human adrenal homogenates. *J. Clin. Invest.* **71**, 933–944.
57. Chester-Jones, I. & Wright, A. (1954). Some aspects of zonation and function of the adrenal cortex. IV. The histology of the adrenal in rats with diabetes insipidus. *J. Endrocrinol.* **10**, 266–272.
58. Muller, J. (1971). In *Regulation of Aldosterone Biosynthesis* (Gross, F., Labhart, A., Mann, T., Samuels, L. T. & Zander, J., eds.). Monographs on Endocrinology, Vol. 5. Springer-Verlag, Berlin, Heidelberg and New York.
59. Marusic, E. T. & Mulrow, P. J. (1967). Stimulation of aldosterone biosynthesis in adrenal mitochondria by sodium depletion *J. Clin. Invest.* **46**, 2101–2108.
60. Blair-West, J. R., Brodie, A., Coghlan, J. P., Denton, D. A., Flood, C., Goding, J. R., Scoggins, B. A., Tait, J. F., Tait, S. A. S., Wintour, E. M. & Wright, R. D. (1970). Studies on the biosynthesis of aldosterone using the sheep adrenal transplant: effect of sodium depletion on the conversion of corticosterone to aldosterone. *J. Endocrinol.* **46**, 453–476.
61. Aguilera, G. & Marusic, E. T. (1971). Role of the renin-angiotensin system in the biosynthesis of aldosterone. *Endocrinology* **89**, 1524–1529.
62. Kramer, R. E., Gallant, S. & Brownie, A. C. (1979). The role of cytochrome P-450 in the action of sodium depletion on aldosterone biosynthesis in rats. *J. Biol. Chem.* **254**, 3953–3958.
63. Kramer, R. E., Gallant, S. & Brownie, A. C. (1980). Actions of angiotensin II on aldosterone biosynthesis in the rat adrenal cortex: effect on cytochrome P-450 enzymes of the early and late pathway. *J. Biol. Chem.* (in the press).
64. Boyd, J. E., Palmore, W. P. & Mulrow, P. J. (1971). Role of potassium in the control of aldosterone secretion in the rat. *Endocrinology* **88**, 556–565.
65. Kaplan, N. M. (1965). The biosynthesis of adrenal steroids: effects of angiotensin II, adrenocorticotropin and potassium. *J. Clin. Invest.* **44**, 2029–2039.
66. Szalay, K. S. (1971). The effect of ouabain on aldosterone production in the rat. *Acta Endocrinol.* (*Copenhagen*) **68**, 477–484.
67. Wellen, J. J. & Benraad, T. J. (1969). Effect of ouabain on corticosterone biosynthesis and on potassium and sodium concentration in calf adrenal tissue *in vitro*. *Biochim. Biophys. Acta* **183**, 100–117.
68. Blaine, E. H., Coghlan, J. P., Denton, D. A. & Scoggins, B. A. (1974). *In vivo* effects of ouabain on aldosterone, corticosterone and cortisol secretion in conscious sheep. *Endocrinology* **94**, 1304–1310.
69. Gill, J. R., Frolich, J. C., Bowden, R. E., Taylor, A. A., Keiser, H. R., Seyberth, H. W., Oates, J. A. & Bartter, F. C. (1976). Bartter's syndrome: a disorder characterized by high urinary prostaglandins and a dependence of hyperreninemia on prostaglandin synthesis. *Am. J. Med.* **61**, 43–51.
70. Campbell, W. B., Gomez-Sanchez, C. E., Adams, B. V., Schmitz, J. M. & Itskovitz, H. D. (1979). Attenuation of angiotensin II- and III-induced aldosterone release by prostaglandin synthesis inhibitors. *J. Clin. Invest.* **64**, 1552–1557.

71. Spat, A. & Jozan, S. (1975). Effect of prostaglandin E_2 and A_2 on steroid synthesis by the rat adrenal gland. *J. Endocrinol.* **65**, 55–63.
72. O'Hare, M. J. & Neville, A. M. (1973). The steroidogenic response of adult rat adrenocortical cells in monolayer culture. *J. Endocrinol.* **56**, 537–549.
73. Hornsby, P. J., O'Hare, M. J. & Neville, A. M. (1977). Functional and morphological observations on rat adrenal zona glomerulosa cells in monolayer culture. *Endocrinology* **95**, 1240–1251.
74. O'Hare, M. J., Nice, E. C. & Neville, A. M. (1980) Regulation of androgen secretion and sulphoconjugation in the human adrenal cortex: studies with primary monolayer cell cultures. In *Adrenal Androgens* (Genazzani, A. R., Thijssen, J. H. H. & Siiteri, P. K., eds.). Raven Press, New York (in the press).
75. Tait, J. F., Tait, S. A. S., Gould, R. P. & Mee, M. S. R. (1974). The properties of adrenal zona glomerulosa cells after purification by gravitational sedimentation. *Proc. R. Soc. London Ser. B.* **185**, 375–407
76. Lowry, P. J. & McMartin, C. (1974). Measurement of the dynamics of stimulation and inhibition of steroidogenesis in isolated rat adrenal cells by using column perfusion. *Biochem. J.* **142**, 287–294.
77. Williams, B. C., McDougall, J. G., Tait, J. F., Tait, S. A. S. & Zananiri, F. A. F. (1980). Calcium efflux from superfused isolated rat adrenal glomerulosa cells. *Abst, Int. Congr. Endocrinol.* VIth Melbourne, Australia, No. 655.
78. Boscaro, M., Al-Dujaili, E. A. S. & Edwards, C. R. W. (1979). *In vitro* studies on the mechanism of inhibition by bromocriptine, lisuride and metergoline of angiotensin II induced aldosterone release. *Acta Endocrinol.* (*Copenhagen*) *Suppl. 225*, **91**, 342.
79. Schulster, D. (1973). Regulation of steroidogenesis by ACTH in a superfusion system for isolated adrenal cells. *Endocrinology* **93**, 700–704.
80. Joseph, T., Slack, C. & Gould, R. P. (1973). Gap junctions and electrotonic coupling in foetal rabbit adrenal cortical cells. *J. Embryol. Exp. Morphol.* **29**, 681–696.
81. Chayen, J., Daly, J. R., Loveridge, N. & Bitensky, L. (1976). The cytochemical bioassay of hormones. *Recent Prog. Horm. Res.* **32**, 33–79.
82. Attwell, D., Cohen, I. & Eisner, D. (1979). Membrane potential and ion concentration stability conditions for a cell with a restricted extracellular space. *Proc. R. Soc. London Ser. B.* **206**, 145–161.
83. Saffran, M. & Schally, A. V. (1955). *In vitro* bioassay of corticotropin: modification and statistical treatment. *Endocrinology* **56**, 523–532.
84. Barofsky, A-L., Feinstein, M. & Halkerston, I. D. K. (1973). Enzymatic and mechanical requirements for the dissociation of cortical cells from rat adrenal glands. *Exp. Cell Res.* **79**, 263–274.
85. Kono, T. (1969). Role of collagenases and other proteolytic enzymes in the dispersal of animal tissues, *Biochim. Biophys. Acta* **17**, 397–400.
86. Kono, T. & Barham, F. W. (1971). Insulin-like effects of trypsin on fat cells. *J. Biol. Chem.* **246**, 6204–6209.
87. Neher, R. & Milani, A. (1978). Steroidogenesis in isolated adrenal cells; excitation by calcium. *Mol. Cell. Endocrinol.* **9**, 243–253.
88. Kloppenborg, P. W. C., Island, D. P., Liddle, G. W., Michelakis, A. M. & Nicholson, W. E. (1968). A method of preparing adrenal cell suspensions and its applicability to the *in vitro* study of adrenal metabolism. *Endocrinology* **82**, 1053–1058.
89. Haning, R., Tait, S. A. S. & Tait, J. F. (1970). *In vitro* effects of ACTH, angiotensins, serotonin and potassium on steroid output and conversion of

corticosterone to aldosterone by isolated adrenal cells. *Endocrinology* **87**, 1147–1167.

90. Williams, G. H., McDonnell, L. M., Raux, M. & Hollenberg, N. K. (1974). Evidence for different angiotensin II receptors in rat adrenal glomerulosa and rabbit vascular smooth muscle cells. *Circ. Res.* **34**, 384–390.

91. Bing, R. F. & Schulster, D. (1977). Steroidogenesis in isolated rat adrenal glomerulosa cells: response to physiological concentrations of angiotensin II and effects of potassium, serotonin and [Sar^1, Ala^8] angiotensin II. *J. Endocrinol.* **74**, 261–272.

92. Swallow, R. L. & Sayers, G. (1969). A technic for the preparation of isolated rat adrenal cells. *Proc. Soc. Exp. Biol. Med.* **131**, 1–4.

93. Lowry, P. J., McMartin, C. & Peters, J. (1973). Properties of a simplified bioassay for adrenocorticotrophic activity using the steroidogenic response of isolated adrenal cells. *J. Endocrinol.* **59**, 43–55.

94. Edwards, C. R. W., Al-Dujaili, E. A. S., Boscaro, M., Quyyumi, S., Miall, P. A. & Rees, L. H. (1980). *In-vivo* and *in-vitro* studies on the effect of metoclopramide on aldosterone secretion. *Clin. Endocrinol.* (in the press).

95. Richardson, M. C. & Schulster, D. (1972). Corticosteroidogenesis in isolated adrenal cells: effect of adrenocorticotrophic hormone, adenosine 3′,5′-mono-phosphate and β^{1-24} adrenocorticotrophic hormone diazotized to polyacrylamide. *J. Endocrinol.* **55**, 127–139.

96. Hopkins, C. R. & Farquhar, M. G. (1973). Hormone secretion by cells dissociated from rat anterior pituitaries. *J. Cell Biol.* **59**, 276–303.

97. Powell, T. & Twist, V. W. (1976). A rapid technique for the isolation and purification of adult cardiac muscle cells having respiratory control and a tolerance to calcium. *Biochem. Biophys. Res. Common.* **72**, 327–333.

98. Schlossman, S. F. & Hudson, L. (1973). Specific purification of lymphocyte populations on a digestible immunoabsorbent. *J. Immunol.* **110**, 313–315.

99. Seglen, P. O. (1979). Disaggregation and separation of rat liver cells. In *Methodological Surveys in Biochemistry* (Reid, E., ed.), Vol. 8, pp. 25–46. Ellis Horwood, Chichester.

100. Loos, J. A. & Roos, D. (1974). Ficoll-Isopaque gradients for the determination of density distributions of human blood lymphocytes and other reticuloendothelial cells. *Exp. Cell Res.* **86**, 333–341.

101. Ungar, F., Hsiao, J., Greene, J. M. & Headon, D. R. (1979). Isolation of ACTH responsive cells from rat adrenal cortex and the determination of the density of male and female cells. *FEBS Lett.* **108**, 331–334.

102. Dufau, M. L., Horner, K. A., Hayashi, K., Tsuruhara, T., Conn, P. M. & Catt, K. J. (1977). Actions of choleragen and gonadotropin in isolated Leydig cells. Functional compartmentalization of the hormone-activated cyclic AMP response. *J. Biol. Chem.* **253**, 3721–3729.

103. Conn, P. M., Tsuruhara, T., Dufau, M. & Catt, K. J. (1977). Isolation of highly purified Leydig cells by density gradient centrifugation. *Endocrinology* **101**, 639–649.

104. Comments on article by Pertoft, H., Hirstenstein, M. & Kagedal, L. (1979). Cell separations in a new density gradient medium, Percoll. In *Methological Surveys in Biochemistry* (Reid, E., ed.), Vol. 8, p. 216. Ellis Horwood, Chichester.

105. Janszen, F. H. A., Cooke, B. A., van Driel, M. J. A. & van der Molen, H. J. (1976). Purification and characterization of Leydig cells from rat testes. *J. Endocrinol.* **70**, 345–359.

106. Miller, R. G. & Phillips, R. A. (1969). Separation of cells by velocity sedimentation. *J. Cell. Physiol.* **73**, 191–201.
107. Zeiller, K., Hansen, E., Leihener, D., Pascher, G. & Wirth, H. (1976). Analysis of velocity sedimentation techniques in cell separation. Influence of apparative and sample properties on separative power, resolution and sensitivity. *Hoppe-Seyler's Z. Physiol. Chem.* **357**, 1309–1319.
108. Braley, L. M. & Williams, G. H. (1980). The effect of unit gravity sedimentation on adrenal steroidogenesis by isolated rat glomerulosa and fasciculata cells. *Endocrinology* **106**, 50–55.
109. Bell, J. B. G., Bhatt, K., Hyatt, P. J., Tait, J. F. & Tait, S. A. S. (1980) Properties of adrenal zona reticularis cells. In *Adrenal Androgens* (Genazzani, A. R., Thijssen, J. H. H. & Siiteri, P. K., eds.). Raven Press, New York (in the press).
110. McDougall, J. G. & Williams, B. C. (1978). An improved method for the superfusion of dispersed rat adrenal cells. *J. Endocrinol.* **78**, 157–158.
111. McDougall, J. G., Williams, B. C., Hyatt, P. J., Bell, J. B. G., Tait, J. F. & Tait, S. A. S. (1979). Purification of dispersed rat adrenal cells by column filtration. *Proc. R. Soc. London Ser. B.* **206**, 15–32.
112. Fujita, K., Aguilera, G. & Catt, K. J. (1979). The role of cyclic AMP in aldosterone production by isolated zona glomerulosa cells. *J. Biol. Chem.* **254**, 8567–8574.
113. Seelig, S. & Sayers, G. (1973). Isolated adrenal cortex cells: ACTH agonists, partial agonists, antagonists; cyclic AMP and corticosteroid production. *Arch. Biochem. Biophys.* **154**, 230–239.
114. Schwyzer, R, (1964). Chemistry and metabolic action of nonsteroid hormones. *Annu. Rev. Biochem.* **33**, 259–286.
115. Shanker, G. & Sharma, R. K. (1979). β-endorphin stimulates corticosterone synthesis in isolated rat adrenal cells. *Biochem. Biophys. Res. Commun.* **86**, 1–5.
116. Tait, J. F., Tait, S. A. S., Bell, J. B. G., Hyatt, P. J. & Williams, B. C. (1980). Further studies on the stimulation of rat adrenal capsular cells. Four types of responses. *J. Endocrinol.* (in the press).
117. Peytremann, A., Nicholson, W. E., Brown, R. D., Liddle, G. W. & Hardman, J. G. (1973). Comparative effects of angiotensin and ACTH on cyclic AMP and steroidogenesis in isolated bovine adrenal cells. *J. Clin. Invest.* **52**, 835–842.
118. Hepp, R., Grillet, A., Peytremann, A. & Valloton, M. B. (1977). Stimulation of corticosteroid biosynthesis by angiotensin I. [des-Asp¹] angiotensin I, angiotensin II and [des-Asp¹] angiotensin II in bovine adrenal fasciculata cells. *Endocrinology* **100**, 717–725.
119. Kaplan, N. M. & Bartter, F. C. (1962). The effect of ACTH, renin, angiotensin II and various precursors on biosynthesis of aldosterone by adrenal slices. *J. Clin. Invest.* **41**, 715–724.
120. Bravo, E. L., Saito, I., Sen, S. & Bumpus, F. M. (1979). Steroidogenic responses to a human urinary protein fraction (UF) by cat adrenocortical collagenase-dispersed cells. *Abst. Endocrine Soc.* 61st California U.S.A., No. 106.
121. Slater, J. D. H., Barbour, B. H., Henderson, H. H., Casper, A. G. T. & Bartter, F. C. (1963). Influence of the pituitary and the renin-angiotensin system on the secretion of aldosterone, cortisol and corticosterone. *J. Clin. Invest.* **42**, 1504–1520.
122. Carpenter, C. C. J., Davis, J. O. & Ayres, C. R. (1961). Relation of renin, angiotensin II and experimental renal hypertension to aldosterone secretion. *J. Clin. Invest.* **40**, 2026–2042.

123. Ames, R. P., Borkowski, A. J., Sicinski, A. M. & Laragh, J. H. (1965). Prolonged infusions of angiotensin II and norepinephrine and blood pressure, electrolyte balance and aldosterone and cortisol secretion in normal man and in cirrhosis with ascites. *J. Clin. Invest.* **44**, 1171–1186.

124. Fraser, R., Mason, P. A., Buckingham, J. C., Gordon, R. D., Morton, J. J., Nicholls, M. G., Semple, P. F. & Tree, M. (1979). The interaction of sodium and potassium status, of ACTH and of angiotensin II in the control of corticosteroid secretion. *J. Steroid Biochem.* **11**, 1039–1042.

125. Horner, J. M., Hintz, R. L. & Luetscher, J. A. (1979). The role of renin and angiotensin in salt-losing, 21-hydroxylase deficient congenital adrenal hyperplasia. *J. Clin. Endocrinol. Metab.* **48**, 776–783.

126. Hepp, R., Grillet, C., Peytremann, A. & Vallotton, M. B. (1976). Stimulating effects of angiotensin I, angiotensin II and des-Asp^1-angiotensin II on steroid production *in vitro* and its inhibition by Sar^1-Ala^8-angiotensin II. *Prog. Biochem. Pharmacol.* **12**, 41–48.

127. Katzen, H. M. & Vlahakes, G. J. (1973). Biological activity of insulin-sepharose? *Science* **179**, 1142–1144.

128. McIlhinney, R. A. J. & Schulster, D. (1975). Studies on the binding of ^{125}-I-labelled corticotrophin to isolated rat adrenocortical cells. *J. Endocrinol.* **64**, 175–184.

129. Bristow, A. F., Gleed, C., Fauchere, J-L., Schwyzer, R. & Schulster, D. (1980). Effects of ACTH (corticotropin) analogues on steroidogenesis and cyclic AMP in rat adrenocortical cells. *Biochem. J.* **186**, 599–603.

130. Limbrey, G. (1979). The binding of adrenocorticotrophic hormone by dispersed and purified cells from the rat adrenal cortex. Ph.D. Thesis, University of London.

131. Catt, K. J., Harwood, J. P., Aguilera, G. & Dufau, M. L. (1979). Hormonal regulation of peptide receptors and target cell responses. *Nature (London)* **280**, 109–116.

132. Aguilera, G., Hauger, R. L. & Catt, K. J. (1978). Control of aldosterone secretion during sodium restriction: adrenal receptor regulation and increased adrenal sensitivity to angiotensin II. *Proc. Natl. Acad. Sci. U.S.A.* **75**, 975–979.

133. Posner, B. I., Kelly, P. A., Shiu, P. C. & Friesen, H. G. (1974) Studies of insulin, growth hormone and prolactin binding: tissue distribution, species variation and characterization. *Endocrinology* **95**, 521–531.

134. Schulster, D., Orly, J., Seidel, G. & Schramm, M. (1978). Intracellular cyclic AMP production enhanced by a hormone receptor transferred from a different cell. *J. Biol. Chem.* **253**, 1201–1206.

135. Dufau, M. L., Hayashi, K., Sala, G., Baukal, A. & Catt, K. J. (1978). Gonadal luteinizing hormone receptors and adenylate cyclase: transfer of functional ovarian luteinizing hormone receptors to adrenal fasciculata cells. *Proc. Natl. Acac. Sci. U.S.A.* **75**, 4769–4773.

136. Beall, R. J. & Sayers, G. (1972). Isolated adrenal cells: steroidogenesis and cyclic AMP accumulation in response to ACTH. *Arch. Biochem. Biophys.* **148**, 70–76.

137. Podesta, E. J., Milani, A., Steffen, H. & Neher, R. (1979). Steroidogenesis in isolated adrenocortical cells. Correlation with receptor-bound adenosine 3',5'-cyclic monophosphate. *Biochem. J.* **180**, 355–363.

138. Mackie, C., Richardson, M. C. & Schulster, D. (1972). Kinetics and dose-response characteristics of adenosine 3',5'-monophosphate production by isolated rat adrenal cells stimulated with adrenocorticotrophic hormone. *FEBS Lett.* **23**, 345–348.

139. Kolanowski, J. & Crabbe, J. (1976). Characteristics of the response of human adrenocortical cells to ACTH. *Mol. Cell. Endocrinol.* **5**, 255–267.
140. Saez, J. M., Evain, D. & Gallet, D. (1978). Role of cyclic AMP and protein kinase on the steroidogenic action of ACTH, prostaglandin E_1 and dibutyryl cyclic AMP in normal adrenal cells and adrenal tumour cells from humans. *J. Cyclic Nucl. Res.* **4**, 311–321.
141. Rommerts, F. F. G., Cooke, B. A. & van der Molen, H. J. (1974). The role of cyclic AMP in the regulation of steroid biosynthesis in testis tissue. *J. Steroid Biochem.* **5**, 279–285.
142. Hudson, A. M. & McMartin, C. (1975). An investigation of the involvement of adenosine 3′:5′-cyclic monophosphate in steroidogenesis by using isolated adrenal cell column perfusion. *Biochem. J.* **148**, 539–544.
143. Espiner, E. A., Livesey, J. H., Ross, J. & Donald, R. A. (1974). Dynamics of cyclic adenosine 3′,5′-monophosphate release during adrenocortical stimulation *in vivo*. *Endocrinology* **95**, 838–846.
144. Dufau, M. L., Tsuruhara, T., Horner, K. A., Podesta, E. & Catt, K. J. (1977). Intermediate role of adenosine 3′:5′-cyclic monophosphate and protein kinase during gonadotropin-induced steroidogenesis in testicular interstitial cells. *Proc. Natl. Acad. Sci. U.S.A.* **74**, 3419–3423.
145. Sala, G. B., Hayashi, K., Catt, K. J. & Dufau, M. L. (1979). Adrenocorticotrophin action in isolated adrenal cells. The intermediate role of cyclic AMP in stimulation of corticosterone synthesis. *J. Biol. Chem.* **254**, 3861–3865.
146. Halkerston, I. D. K. (1968). Heterogeneity of the response of adrenal cortex tissue slices to adrenocorticotrophin. In *Functions of the Adrenal Cortex* (McKerns, K. W., ed.), Vol. 1, pp. 399–461. Appleton-Century-Crofts, New York.
147. Chapman, R. A. (1979). Excitation-contraction coupling in cardiac muscle. *Prog. Biophys. Molec. Biol.* **35**, 1–52.
148. Cooke, B. A., Lindh, M. L. & Janszen, F. H. A. (1976). Correlation of protein kinase activation and testosterone production after stimulation of Leydig cells with luteinizing hormone. *Biochem. J.* **160**, 439–446.
149. Harper, J. F. & Brooker, G. (1975). Femtomole sensitive radioimmunoassay for cyclic AMP and cyclic GMP after 2′O-acetylation by acetic anhydride in aqueous solution. *J. Cyclic Nucl. Res.* **1**, 207–218.
150. Hyatt, P. J., Wale, L. W., Bell, J. B. G., Tait, J. F. & Tait, S. A. S. (1980). Adenosine 3′:5′ cyclic monophosphate levels in purified rat adrenal zonae fasciculata and reticularis cells and the effects of adrenocorticotrophic hormone. *J. Endocrinol.* **85**, 435–442.
151. Moyle, W. R., Kong, Y. C. & Ramachandran, J. (1973). Steroidogenesis and cyclic adenosine 3′,5′-monophosphate accumulation in rat adrenal cells. *J. Biol. Chem.* **248**, 2409–2417.
152. Laychock, S. G., Warner, W. & Rubin, R. P. (1977). Further studies on the mechanisms controlling prostaglandin biosynthesis in the cat adrenal cortex: the role of calcium and cyclic AMP. *Endocrinology* **100**, 74–81.
153. Richardson, M. C. & Schulster, D. (1973). The role of protein kinase activation in the control of steroidogenesis by adrenocorticotrophic hormone in the adrenal cortex. *Biochem. J.* **136**, 993–998.
154. Podesta, E. J., Dufau, M. L., Solano, A. R. & Catt, K. J. (1978), Hormonal activation of protein kinase in isolated Leydig cells. *J. Biol. Chem.* **253**, 8994–9001.

155. Podesta, E. J., Milani, A., Steffen, H. & Neher, R. (1979). Adrenocorticotrophin (ACTH) induces phosphorylation of a cytoplasmic protein in intact isolated adrenocortical cells. *Proc. Natl. Acad. Sci. U.S.A.* **76**, 5187–5191.

156. Sharma, R. K., Ahmed, N. K., Sutliff, L. S. & Brush, J. S. (1974). Metabolic regulation of steroidogenesis in isolated adrenal cells of the rat. ACTH regulation of cGMP and cAMP levels and steroidogenesis. *FEBS. Lett.* **45**, 107–110.

157. Rodbell, M. (1980). The role of hormone receptors and GTP-regulatory proteins in membrane transduction. *Nature (London)* **284**, 17–21.

158. Perchellet, J. P. & Sharma, R. K. (1979). Mediatory role of calcium and guanosine 3′,5′-monophosphate in adrenocorticotropin-induced steroidogenesis by adrenal cells. *Science* **20**, 1259–1261.

159. Michell, R. H. (1979). Inositol phospholipids in membrane function. *Trends Biochem. Sci.* **4**, 128–131.

160. Halkerston, I. D. K. (1975). Cyclic AMP and adrenocortical function. *Adv. Cyclic Nucl. Res.* **6**, 99–136.

161. Shima, S., Kawashima, Y. & Hirai, M. (1978). Studies on cyclic nucleotides on the adrenal gland. VIII. Effects of angiotensin on adenosine 3′,5′-monophosphate and steroidogenesis in the adrenal cortex. *Endocrinology* **103**, 1361–1367.

162. Fakunding, J. L. (1979). The role of calcium in aldosterone production by isolated adrenal glomerulosa cells. *Abst. Endocrine Soc.* 61st California, U.S.A., No. 103.

163. Berridge, M. J. (1975). Calcium and cyclic nucleotides. *Adv. Cyclic Nucl. Res.* **6**, 1–98.

164. Mackie, C., Warren, R. L. & Simpson, E. R. (1978). Investigations into the role of calcium ions in the control of steroid production by isolated adrenal zona glomerulosa cells of the rat. *J. Endocrinol.* **77**, 119–127.

165. Borle, A. B. (1972). Kinetic analysis of calcium movements in cell culture. V. Intracellular calcium distribution in kidney cells. *J. Membrane Biol.* **10**, 45–66.

166. Rasmussen, H. & Goodman, D. B. P. (1977). Relationships between calcium and cyclic nucleotides in cell activation. *Physiol. Rev.* **57**, 421–509.

167. Rasmussen, H. & Gustin, M. C. (1978). Some aspects of the hormonal control of cellular calcium metabolism. *Ann. N.Y. Acad. Sci.* **307**, 391–401.

168. Sayers, G., Beall, R. J. & Seelig, S. (1972). Isolated adrenal cells: adrenocorticotrophic hormone, calcium, steroidogenesis and cyclic adenosine monophosphate. *Science* **175**, 1131–1132.

169. Haksar, A. & Peron, F. G. (1973). The role of calcium in the steroidogenic response of rat adrenal cells to adrenocorticotropic hormone. *Biochim. Biophys. Acta* **313**, 363–371.

170. Birmingham, M. K. & Bartova, A. (1973). Effects of calcium and theophylline on ACTH- and dibutyryl cyclic AMP-stimulated steroidogenesis and glycolysis by intact mouse adrenal glands *in vitro. Endocrinology* **92**, 743–749.

171. Bar, H. P. & Hechter, O. (1969). Adenyl cyclase and hormone action. III. Calcium requirement for ACTH stimulation of adenyl cyclase. *Biochem. Biophys. Res. Commun.* **35**, 681–686.

172. Lefkowitz, R. J., Roth, J. & Pastan, I. (1971). ACTH-receptor interaction in the adrenal. A model for the initial step in the action of hormones that stimulate adenyl cyclase. *Ann. N.Y. Acad. Sci.* **185**, 195–209.

173. Simpson, E. R. & Williams-Smith, D. L. (1975). Effect of calcium (ion) uptake by rat adrenal mitochondria on pregnenolone formation and spectral properties of cytochrome P-450. *Biochim. Biophys. Acta* **404**, 309–320.

174. Farese, R. V. (1971). Stimulatory effects of calcium on protein synthesis on adrenal (and thyroidal) cell-free systems as related to trophic hormone action. *Endocrinology* **89**, 1064–1074.

175. Leier, D. J. & Jungmann, R. A. (1973). Adrenocorticotropic hormone and dibutyryl adenosine cyclic monophosphate-mediated Ca^{2+} uptake by rat adrenal glands. *Biochim. Biophys. Acta* **329**, 196–210.

176. Jaanus, S. D. & Rubin, R. P. (1971). The effect of ACTH on calcium distribution in the perfused cat adrenal gland. *J. Physiol. (London)* **213**, 581–598.

177. Mathews, E. K. & Saffran, M. (1973). Ionic dependence of adrenal steroidogenesis and ACTH-induced changes in the membrane potential of adrenocortical cells. *J. Physiol. (London)* **234**, 43–64.

178. Rubin, R. P., Sheid, B., McCauley, R. & Laychock, S. G. (1974). ACTH induced protein release from perfused cat adrenal gland; evidence for exocytosis. *Endocrinology* **96**, 370–378.

179. Gemmell, R. T., Laychock, S. G. & Rubin, R. P. (1977). Ultrastructural and biochemical evidence for steroid-containing secretory organelle in the perfused cat adrenal gland. *J. Cell. Biol.* **72**, 209–215.

180. Nussdorfer, G. G., Mazzocchi, G. & Meneghelli, V. (1978). Cytophysiology of the adrenal zona fasciculata. *Int. Rev. Cytol.* **55**, 291–365.

181. Ray, P. & Strott, C. A. (1978). Stimulation of steroid synthesis by normal rat adrenocortical cells in response to antimicrotubular agents. *Endocrinology* **103**, 1281–1288.

182. Goddard, C., Vinson, G. P. & Whitehouse, B. J. (1978). Steroid and protein synthesis and secretion by rat adrenocortical tissue *in vitro*. *J. Endocrinol.* **77**, 10P–11P.

183. Podesta, E. J., Milani, A., Steffen, H. & Neher, R. (1980). Steroidogenic action of calcium ions in isolated adrenocortical cells. *Biochem. J.* **186**, 391–397.

184. Neher, R. & Milani, A. (1976). Mode of action of peptide hormones. *Clin. Endocrinol. Suppl.* **5**, 29s–39s.

185. Neher, R. & Milani, A. (1978). Steroidogenesis in isolated adrenal cells: excitation by calcium. *Mol. Cell. Endocrinol.* **9**, 243–253.

186. Rimon, G., Hanski, E., Braun, S. & Levitzki, A. (1978). Mode of coupling between hormone receptors and adenylate cyclase elucidated by modulation of membrane fluidity. *Nature (London)* **276**, 394–396.

187. Rubin, R. P. & Laychock, S. G. (1978). Prostaglandins and calcium-membrane interaction in secretory glands. *Ann. N.Y. Acad. Sci.* **307**, 377–390.

188. Flack, J. D., Jessup, R. & Ramwell, P. W. (1969). Prostaglandin stimulation of rat corticosteroidogenesis. *Science* **163**, 691–692.

189. Catt, K. J., Anguilera, G., Capponi, A., Fujita, K., Saruta, A. & Fakunding, J. (1979). Angiotensin II receptors and aldosterone secretion. *J. Endocrinol.* **81**, 37P–48P.

190. Nakajima, T., Khosla, M. C. & Sakakibara, S. (1978). Comparative biochemistry of renins and angiotensins in the vertebrates. *Japanese Heart J.* **19**, 799–805.

191. Peytremann, A., Brown, R. D., Nicholson, W. E., Island, D. P., Liddle, G.W. & Hardman, J. G. (1974). Regulation of aldosterone synthesis. *Steroids* **24**, 451–462.

192. Braley, L. M. & Williams, G. H. (1977). Rat adrenal cell sensitivity to angiotensin II, a^{1-24} ACTH and potassium: a comparative study. *Am. J. Physiol.* **233**, E402–E406.

193. Fredlund, P., Saltman, S. & Catt, K. J. (1975). Aldosterone production by isolated adrenal glomerulosa cells: stimulation by physiological concentrations of angiotensin II. *Endocrinology* **97**, 1577–1586.
194. Lobo, M. V., Marusic, E. T. & Aguilera, G. (1978). Further studies on the relationship between potassium and sodium levels and adrenocortical activity. *Endocrinology* **102**, 1061–1068.
195. Braley, L. M. & Williams, G. H. (1978). The effects of ouabain on steroid production by rat adrenal cells stimulated by angiotensin II, α^{1-24} adrenocorticotropin and potassium. *Endocrinology* **103**, 1997–2005.
196. Peach, M. J., Sarstedt, C. A. & Vaughan, E. D. (1976). Changes in cardiovascular and adrenal cortical responses to angiotensin III induced by sodium deprivation in the rat. *Circ. Res. Suppl. II* to **38**, 117–121.
197. Semple, P. F. & Morton, J. J. (1976). Angiotensin II and angiotensin III in rat blood. *Circ. Res. Suppl. II* to **38**, 122–126.
198. Blair-West, J. R., Coghlan, J. P., Denton, D. A., Funder, J. W., Scoggins, B. A. & Wright, R. D. (1971). The effect of the heptapeptide (2-8) and hexapeptide (3-8) fragments of angiotensin II on aldosterone secretion. *J. Clin. Endocrinol. Metab.* **32**, 575–578.
199. Goodfriend, T. L. & Peach, M. J. (1975). Angiotensin III: (Des-aspartic acid^1)-angiotensin II. Evidence and speculation for its role as an important agonist in the renin-angiotensin system. *Circ. Res. Suppl. I* to **36** and **37**, 38–48.
200. Aguilera, G., Capponi, A., Baukal, A., Fujita, K., Hauger, R. & Catt, K. J. (1979). Metabolism and biological activities of angiotensin II and des-Asp^1-angiotensin II in isolated adrenal glomerulosa cells. *Endocrinology* **104**, 1279–1285.
201. Douglas, J. G., Michailov, M., Khosla, M. C. & Bumpus, F. M. (1979). Comparative studies of receptor binding and steroidogenic properties of angiotensins in the rat adrenal glomerulosa. *Endocrinology* **104**, 71–75.
202. Fei, D. T. W., Coghlan, J. P., Fernley, R. T. & Scoggins, B. A. (1979). Blood clearance rates of angiotensin II and its metabolites in sheep: presence of immunoreactive fragments in arterial blood. *Clin. Exp. Pharmacol. Physiol.* **6**, 129–137.
203. Kharaillah, P. A., Khosla, M. C., Bumpus, F. M., Tait, J. F. & Tait, S. A. S. (1978). Steroidogenic and pressor activity of angiotensin analogues in the rat. *Clin. Sci. Mol. Med.* **55**, 175s–177s.
204. Fredlund, P., Saltman, S. & Catt, K. J. (1975). Stimulation of aldosterone production by angiotensin II peptides *in vitro*: enhanced activity of the (1-sarcosine) analogue. *J. Clin. Endocrinol. Metab.* **40**, 746–749.
205. Peach, M. J. & Chiu, A. T. (1974). Stimulation and inhibition of aldosterone biosynthesis *in vitro* by angiotensin II and analogues. *Circ. Res. Suppl. I* to **34** and **35**, 7–13.
206. Saltman, S., Fredlund, P. & Catt, K. J. (1976). Action of angiotensin II antagonists upon aldosterone production by isolated adrenal glomerulosa cells. *Endocrinology* **98**, 894–903.
207. Rodbell, M. (1964). Metabolism of isolated fat cells. I. Effects of hormones on glucose metabolism and lipolysis. *J. Biol. Chem.* **239**, 375–380.
208. Norbiato, G., Berilacqua, M. Raggi, U., Micossi, P. & Moroni, G. (1977). Metoclopramide increases plasma aldosterone concentration in man. *J. Clin. Endocrinol. Metab.* **45**, 1313–1316.
209. McKenna, T. J., Island, D. P., Nicholson, W. E. & Liddle, G. W. (1979). Dopamine inhibits angiotensin stimulated aldosterone biosynthesis in bovine adrenal cells. *J. Clin. Invest.* **64**, 287–291.

210. McKenna, T. J., Island, D. P., Nicholson, W. E., Miller, R. B. & Liddle, G. W. (1978). The role of dopamine in aldosterone biosynthesis. *Clin. Res.* **26**, 16.
211. Brown, R. D., Kao, P. & Jiang, N. S. (1979). Metoclopramide acts directly on the adrenal cortex to stimulate the secretion of aldosterone. *Clin. Res.* **27**, 248A.
212. Saruta, T. & Kaplan, N. M. (1972). Adrenocortical steroidogenesis: the effects of prostaglandins. *J. Clin. Invest.* **51**, 2246–2251.
213. Blair-West, J. R., Coghlan, J. P., Denton, D. A., Funder, J. W., Scoggins, B. A. & Wright, R. D. (1971). Effects of prostaglandin E₁ upon the steroid secretion of the adrenal of the sodium deficient sheep. *Endocrinology* **88**, 367–371.
214. Ferreira, S. H. & Vane, J. R. (1967). Prostaglandins: their disappearance from and release into the circulation. *Nature (London)* **216**, 868–873.
215. Spat, A., Siklos, P., Antoni, F. A., Nagy, K. & Sziranyi, K. (1977). Effect of prostaglandin synthetase inhibitors on basal and ACTH-stimulated steroid synthesis by separated adrenocortical zones. *J. Steroid Biochem.* **8**, 293–298.
216. Spat, A., Jozan, S., Gaal, K. & Mozes, T. (1977). Effect of indomethacin on the adrenal response to frusemide in the rat. *J. Endocrinol.* **73**, 401–402.
217. Patak, R. V., Mookerjee, B. K., Bentzel, C. J., Hysert, P. E., Babej, M. & Lee, J. B. (1975). Antagonism of the effects of furosemide by indomethacin in normal and hypertensive man. *Prostaglandins* **10**, 649–659.
218. McKenna, T. J., Island, D. P., Nicholson, W. E. & Liddle, G. W. (1980). Stimulation of aldosterone production by catecholamines *in vitro*. *Abst. Int. Congr. Endocrinol.* VIth Melbourne, Australia, No. 395.
219. Edwards, C. R. W., Boscaro, M., Miall, P. A., Delitala, G. & Al-Dujaili, E. A. S. (1980). Inhibition of aldosterone secretion by histamine-H₂ receptor antagonists. *Abst. Int. Congr. Endocrinol.* VIth Melbourne, Australia, No. 592.
220. McKenna, T. J., Island, D. P., Nicholson, W. E. & Liddle, G. W. (1978). Angiotensin stimulates both early and late steps in aldosterone biosynthesis in isolated bovine glomerulosa cells. *J. Steroid Biochem.* **9**, 967–972.
221. McKenna, T. J., Island, D. P., Nicholson, W. E. & Liddle, G. W. (1978). The effect of potassium on early and late steps in aldosterone biosynthesis in cells of the zona glomerulosa. *Endocrinology* **103**, 1411–1415.
222. Aguilera, G. & Catt, K. J. (1979). Loci of action of regulators of aldosterone biosynthesis in isolated glomerulosa cells. *Endocrinology* **104**, 1046–1052.
223. Williams, G. H., McDonnell, L. M., Tait, S. A. S. & Tait, J. F. (1972). The effect of medium composition and *in vitro* stimuli on the conversion of corticosterone to aldosterone in rat glomerulosa tissue. *Endocrinology* **91**, 948–958.
224. Brecher, P. I., Pyun, H. Y. & Chobanian, A. V. (1974). Studies on the angiotensin II receptor in the zona glomerulosa of the rat adrenal gland. *Endocrinology* **95**, 1026–1033.
225. Capponi, A. M. & Catt, K. J. (1979). Angiotensin II receptors in adrenal cortex and uterus. Binding and activation properties of angiotensin analogues. *J. Biol. Chem.* **254**, 5120–5127.
226. Douglas, J., Aguilera, G., Kondo, T. & Catt, K. J. (1978). Angiotensin II receptors and aldosterone production in rat adrenal gomerulosa cells. *Endocrinology* **102**, 685–696.
227. Douglas, J., Saltman, S., Fredlund, P., Kondo, T. & Catt, K. J. (1976). Receptor binding of angiotensin II and antagonists. Correlation with aldosterone production by isolated canine adrenal glomerulosa cells. *Circ. Res. Suppl. II* to **38**, 108–111.
228. Gurchinoff, S. & Khairallah, P. A. (1977). Inhibition of ³H-angiotensin II binding to zona glomerulosa cells by several analogues. *Arch. Int. Pharmacodyn.* **228**, 15–22.

229. Regoli, D. (1979). Receptors for angiotensin: a critical analysis. *Canad. J. Physiol. Pharmacol.* **57**, 129–139.
230. Douglas, J. & Catt, K. J. (1976). Regulation of angiotensin II receptors in the rat adrenal cortex by dietary electrolytes. *J. Clin. Invest.* **58**, 834–843.
231. Hauger, R. L., Aguilera, G. & Catt, K. J. (1978). Angiotensin II regulates its receptor sites in the adrenal glomerulosa zone. *Nature (London)* **271**, 176–178.
232. Aguilera, G. & Catt, K. J. (1978). Regulation of aldosterone secretion by the renin-angiotensin system during sodium restriction in rats. *Proc. Natl. Acad. Sci. U.S.A.* **75**, 4057–4061.
233. Devynck, M.-A., Rouzaire-Dubois, B., Chevillotte, E. & Meyer, P. (1976). Variations in the number of uterine angiotensin receptors following changes in plasma angiotensin levels. *Eur. J. Pharmacol.* **40**, 24–37.
234. Pernollet, M-G., Devynck, M-A., Mathews, P. G. & Meyer, P. (1977). Post-nephrectomy changes in adrenal angiotensin II receptors in the rat: influence of exogenous angiotensin and competitive inhibitor. *Eur. J. Pharmacol.* **43**, 361–372.
235. Aguilera, G., Schirar, A., Baukal, A. & Catt, K. J. (1979). Regulation of angiotensin II receptors in the adrenal glomerulosa in the rat. *Abst. Endocrine Soc.* 61st California, U.S.A., No. 824.
236. Douglas, J. G. (1979). Changes in potassium balance: inverse relationship between number and affinity of angiotensin II receptors of smooth muscle and adrenal target tissues. *Am. J. Physiol.* **237**, E519–E523.
237. Coghlan, J. P., Scoggins, B. A. & Wintour, E. M. (1979). Aldosterone. In *Hormones in Blood* (Gray, C. H. & James, V. H. T., eds.), Vol. 3, pp. 493–609. Academic Press, London and New York.
238. Albano, J. D. M., Brown, B. L., Elkins, R. P., Tait, S. A. S. & Tait, J. F. (1974). The effects of potassium, 5-hydroxytryptamine, adrenocorticotrophin and angiotensin II on the concentrations of adenosine 3':5'-cyclic monophosphate in suspensions of dispersed rat adrenal zona glomerulosa and zona fasciculata cells. *Biochem. J.* **142**, 391–400.
239. Tait, S. A. S., Tait, J. F., Gould, R. P., Brown, B. L. & Albano, J. D. M. (1974). The preparation and use of purified and unpurified dispersed adrenal cells and a study of the relationship of their cAMP and steroid output. *J. Steroid Biochem.* **5**, 775–787.
240. Douglas, J., Saltman, S., Williams, C., Bartley, P., Kondo, T. & Catt, K. J. (1978). An examination of possible mechanisms of angiotensin II-stimulated steroidogenesis. *Endocrine Res. Commun.* **5**, 173–188.
241. Saruta, T., Cook, R. & Kaplan, N. M. (1972). Adrenocortical steroidogenesis: studies on the mechanism of action of angiotensin and electrolytes. *J. Clin. Invest.* **51**, 2239–2245.
242. Bing, R. F. & Schulster, D. (1978). Adenosine 3'-5'-cyclic monophosphate production and steroidogenesis by isolated rat adrenal glomerulosa cells. Effects of angiotensin II and [Sar^{1}Ala^8]angiotensin II. *Biochem. J.* **176**, 39–45.
243. Vallotton, M. B., Grillet, C., Knupfer, A. L., Hepp, R., Khosla, M. C. & Bumpus, F. M. (1980). Characterization on angiotensin receptors on bovine adrenal fasciculata cells. *Proc. Natl. Acad. Sci. U.S.A.* (in the press).
244. Fredlund, P., Saltman, S., Kondo, T., Douglas, J. & Catt, K. J. (1977). Aldosterone production by isolated glomerulosa cells: modulation of sensitivity to angiotensin II and ACTH by extracellular potassium concentration. *Endocrinology* **100**, 481–486.

245. Birkhauser, M., Gaillard, R., Riondel, A. M., Scholer, D., Vallotton, M. B. & Muller, A. F. (1973). Effect of volume expansion by hyperosmolar and hyperoncotic solutions under constant infusion of angiotensin II on plasma aldosterone in man and its counterbalance by potassium administration. *Eur. J. Clin. Invest.* **3**, 307–316.

246. Baumber, J. S., Davis, J. O., Johnson, J. A. & Witty, R. T. (1971). Increased adrenocortical potassium in association with increased biosynthesis of aldosterone. *Am. J. Physiol.* **220**, 1094–1099.

247. Boyd, J., Mulrow, P. J., Palmore, W. P. & Silvo. P. (1973). Importance of potassium in the regulation of aldosterone production. *Circ. Res. Suppl. I* to **32** and **33**, 39–45.

248. Mendelsohn, F. A. O., Mackie, C. & Mee, M. S. R. (1975). Measurement of intracellular potassium in dispersed adrenal cortical cells. *J. Steroid Biochem.* **6**, 377–382.

249. Mackie, C. M., Simpson, E. R., Mee, M. S. R., Tait, S. A. S. & Tait, J. F. (1977). Intracellular potassium and steroidogenesis of isolated rat adrenal cells: effect of potassium ions and angiotensin II on purified zona glomerulosa cells. *Clin. Sci. Mol. Med.* **53**, 289–296.

250. Decorzant, C., Riondel, A. M., Philippe, M-J., Bertrand, J. & Vallotton, M. B. (1977). Detection of Na^+ and K^+ in the rat adrenal cortex with the electron microprobe. *Clin. Sci. Mol. Med.* **53**, 423–430.

251. Szalay, K. Sz., Bacsy, E. & Stark, E. (1975). Adrenal potassium and sodium in experimental hyper- and hypoaldosteronism in the rat. Determination by electron probe X-ray microanalysis. *Acta Endocrinol. (Copenhagen)* **80**, 114–125.

252. Foster, R., Lobo, M. V. & Marusic, E. T. (1979). Studies of relationship between angiotensin II and potassium ions on aldosterone release. *Am. J. Physiol.* **237**, E363–E366.

253. Cushman, P. (1969). Inhibition of aldosterone secretion by ouabain in dog adrenal cortical tissue. *Endocrinology* **84**, 808–813.

254. Szalay. K. Sz. (1973). *In vitro* aldosterone production: effect of ethacrynic acid, chlorpromazine and veratrine. *Acta Physiol. Acad. Sci. Hungaricae* **43**, 275–279.

255. Natke, E. & Kabela, E. (1979). Electrical responses in cat adrenal cortex: possible relation to aldosterone secretion. *Am. J. Physiol.* **237**, E158–E162.

256. Fakunding, J. L., Chow, R. & Catt, K. J. (1979). The role of calcium in the stimulation of aldosterone production by adrenocorticotropin, angiotensin II, and potassium in isolated glomerulosa cells. *Endocrinology* **105**, 327–333.

257. Langer, G. A, (1978). The structure and function of the myocardial cell surface. *Am. J. Physiol.* **235**, H461–H468.

258. Bell, J. B. G., Gould, R. P., Hyatt, P. J., Tait, J. F. & Tait, S. A. S. (1979). Properties of rat adrenal zona reticularis cells: production and stimulation of certain steroids. *J. Endocrinol.* **83**, 435–447.

259. Bell, J. B. G., Hyatt, P. J., McDougall, J.G., Tait, J. F., Tait, S. A. S. & Williams, B. C. (1979). Properties of dispersed cells from rat adrenals. *J. Steroid Biochem.* **11**, 169–174.

260. Bourne, G. & Zuckerman, S. (1941). Changes in the adrenals in relation to the normal and artificial threshold oestrous cycle in the rat. *J. Endocrinol.* **2**, 283–310.

261. James, V. H. T., Jones, D. & Jacobs, H. S. (1980). The relationship between plasma prolactin and dehydroepiandrosterone sulphate levels in patients with hyperprolactinaemia. In *Adrenal Androgens* (Genazzani, A. R., Thijssen, J. H. H. & Siiteri, P. K., eds.), Raven Press, New York (in the press).

262. Maroulis, G. B., Sherman, B. & Chapler, T. (1980). Prolactin (PRL) and adrenal function. *Abst. Int. Congr. Endocrinol.* VIth Melbourne, Australia, No. 855.
263. Petersen, O. H. (1980). Role of intra- and extracellular calcium in receptor-mediated ion permeability changes in exocrine gland cells. In *Drug receptors and Their Effectors.* (Birdsall, N. J. M., ed.). Macmillan, London (in the press).
264. d'Auriac, G. A., Baudouin, M. & Meyer, P. (1972). Mechanism of action of angiotensin in smooth muscle cell. Biochemical changes following interaction of the hormone with its membrane receptors. *Circ. Res. Suppl. II* to **30** and **31**, 151–157.
265. Ackerly, J. A., Moore, A. F. & Peach, M. J. (1977). Demonstration of different contractile mechanisms for angiotensin II and des-Asp[1]-angiotensin II in rabbit aortic strips. *Proc. Natl. Acad. Sci. U.S.A.* **74**, 5795–5728.
266. Ichikawa, I., Miele, J. F. & Brenner, B. M. (1979). Reversal of renal cortical actions of angiotensin II by verapamil and manganese. *Kidney Int.* **16**, 137–147.
267. Freer, R. J. (1975). Calcium and angiotensin tachyphylaxis in rat uterine smooth muscle. *Am. J. Physiol.* **228**, 1423–1430.
268. Garrison, J. C., Borland, M. K., Florio, V. A. & Twible, D. A. (1979). The role of calcium ion as a mediator of the effects of angiotensin II, catecholamines and vasopressin on the phosphorylation and activity of enzymes in isolated hepatocytes. *J. Biol. Chem.* **254**, 7147–7156.
269. Cohen, P., Burchell, A., Foulkes, J. G., Cohen, P. T. W., Vanaman, T. C. & Nairn, A. C. (1978). Identification of the Ca^{2+}-dependent modulator protein as the fourth subunit of rabbit skeletal muscle phosphorylase kinase. *FEBS Lett.* **92**, 287–293.
270. DePaoli-Roach, A. A., Roach, P. J. & Larner, J. (1979). Multiple phosphorylation of rabbit skeletal muscle glycogen synthase. *J. Biol. Chem.* **254**, 12062–12068.
271. Hems, D. A., Rodrigues, L. M. & Whitton, P. D. (1978). Rapid stimulation by vasopressin, oxytocin and angiotensin II of glycogen degradation in hepatocyte suspensions. *Biochem. J.* **172**, 311–317.
272. Robison, G. A. & Sutherland, E. W. (1970). Sympathin E, sympathin I, and the intracellular level of cyclic AMP. *Circ. Res. Suppl. I* to **26** and **27**, 147–161.
273. Tait, J. F., Bell, J. B. G., Hyatt, P. J., Tait, S. A. S. & Williams, B. C. (1980). Dispersed cells of the adrenal cortex. *Proc. XXVIII Int. Congr. Physiol. Sci.* **8**, Pergamon Press (in the press).

Subject Index